Archie Architecture

by

Elizabeth Sandoval

Illustrations by

Elizabeth Sandoval
Toby de la Torre
Young Mi Chi

The Salvelinus, The Sockeye, and the Egg-Sucking Leech

Abundance and Diversity in the Bristol Bay Drainage (from the Eyes of an Angler)

Heartstreams Series

I

Trout in the Desert: On Fly Fishing, Human Habits, and the Cold Waters of the Arid Southwest

2015

II

A Tale of Three Rivers: Of Wooly Buggers, Bowling Balls, Cigarette Butts, and the Future of Appalachian Brook Trout

2018

III

A Fine-Spotted Trout on Corral Creek: On the Cutthroat Competition of Native Trout in the Northern Rockies

2021

IV

The Salvelinus, The Sockeye, and the Egg-Sucking Leech: Abundance and Diversity in the Bristol Bay Drainage (from the Eyes of an Angler)

2023

The Salvelinus, The Sockeye, and the Egg-Sucking Leech

Abundance and Diversity in the Bristol Bay Drainage (from the Eyes of an Angler)

Matthew Dickerson

Heartstreams Series, Vol. IV

San Antonio, Texas
2023

The Salvelinus, The Sockeye, and the Egg-Sucking Leech: Abundance and Diversity in the Bristol Bay Drainage (from the Eyes of an Angler)

Photographs (including cover photo) by Glen Alsworth Jr.
 Cover design by Peter Dickerson.

First Edition
ISBN: 978-1-60940-623-3

E-books:
ISBN: 978-1-60940-624-0

Wings Press

Wings Press books are distributed to the trade by
Independent Publishers Group
www.ipgbook.com

All author royalties for this book are donated to The Alsworth Foundation to support at-risk youth in the Bristol Bay area.

Contents

Part III.
Education, Economies, and an Epilogue

Release

In memory of Theodore Bradford Dickerson
(1988-2019)

So, what if I'm fishing?
What if I'm fly fishing,
and it's catch-and-release?
What if I hook a beautiful fish,
full of life,
and it leaps and dances
across the surface of this river?

And what, then, if it breaks the line –
that fine strand that can seem so strong
and be so tenuous –
and drops
with a great splash back down,
down into the mysterious depths?

Then the net is empty
because the fish is gone.
How can I release the big emptiness
from the net, let *it* swim off
into the deep?

Can I let go of nothing?
A hole, a hollow, a space where
the fish should have been?
What if it isn't a net now empty,
but a heart—
my own heart?

Acknowledgments

From Matthew Dickerson

I am thankful to *The Farm Lodge* and *Lake Clark Air* for several years of hospitality, and especially to Glen Alsworth Jr. for friendship, photos, wonderful fishing experiences, and many years of safe flying, including berry-picking excursions while waiting for the cloud ceiling to rise. I am also grateful to Middlebury College for supporting the research and writing of this book over several years, to Alaska State Parks for a 2022 artist residency that provided important time and place for working on this book, and to the Vermont Arts Council for support of that residency (and thus also the writing of this book).

My gratitude is also extended to: my talented, thoughtful, delightful daughter-in-law McKenna for collaborating with me on my 2022 artist residency in Alaska; Branden Hummel for leading me to numerous large northern pike in Lake Clark, the use of a boat and ATV, sharing his fly rod with me on a fly-out fishing trip (when mine got left behind), and especially for friendship; biologists Kurt Fausch (from whom I learned much about stream ecology and Dolly Varden char) and Daniel Schindler (who taught me a great deal about sockeye salmon and landform diversity in Bristol Bay while disabusing me of wrong ideas about salmon-borne nitrogen in the soil); John Branson for excellent historical research in and around Lake Clark National Park, and for friendship and hikes; Don Welty for my first trip to the lakes north of Lake Clark (and my first flight squeezing into a Super Cub); Ranger Alison Eskelin for her important work helping fulfill the mission of Wood-Tikchik State Park (on a very small budget), and for sharing thoughts and insights into that park; Dennis Dauble for helpful comments on the almost-final draft of this book; the wonderful Middlebury College students in my unforgettable 2015 Alaska class; David O'Hara and Rob Green for letting me co-teach with them in Alaska in 2021 (in another unforgettable class), and all the wonderful Augustana University students who joined us; my friends Rich, Rick, and Phil, my sons and daughters-in-law Mark, Ellie, Peter, and McKenna, and my nephew Michael for joining me on research trips to the Bristol Bay watershed; and both *Lake Clark National Park and Preserve* and *Katmai National Park and Preserve* for helping steward and protect the vital headwaters of Bristol Bay.

Most especially I am thankful to my wife Deborah Dickerson for supporting me in all my research travels over several years working on this book, for tending the garden and pets while I was away, and for joining me on a couple of those trips which made them even more enjoyable; and to God who created rivers, mountains, char and salmon, who created the light that makes possible the colors of a Dolly Varden, and who gave me the eyes to see them as the work of a loving, artistic, creator.

From Glen Alsworth, Jr.

I know that I am blessed beyond measure. Most people will never get to experience the perfect mixture of unique opportunities, undeserved resources, and unwavering support that have somehow come together to shape my life here in remote Alaska.

I am thankful for my friend Matthew Dickerson who invited me to be involved in this book. There are some people who become friends due to proximity, but Matthew lives on the other side of America. There are some people who become friends out of necessity, but I would choose to spend time on the river with Matthew on purpose. There are those few people who become friends simply because it was meant to be, and Matthew is one of those few. We may spend a lot of time flying in little airplanes together, wading half-frozen through tundra streams, and throwing flies at the water in hopes of enticing some mystery up out of the deep; but honestly we do all that because we enjoy the fellowship. Flying, wading, exploring, fishing…those are just great ways to hang out together.

I am thankful for my parents Glen Alsworth, Sr. and Patty Alsworth for their patient faith and their constant support, to my grandparents Laddy and Glenda Elliott for being a living example of steadfast commitment, and to my grandparents Leon "Babe" and Mary Alsworth for their pioneering spirits. And I am thankful for the legacy of hard work, respect for others, and the awe of creation that was both displayed and taught to me.

I am eternally thankful to the God who created a world of wonder. And I am thankful to the Creator who gives life and breath, and who has allowed me to live my life exploring one of the most beautiful and rugged parts of creation and breathing the crisp Alaska air.

I am thankful for my five amazing children, Salina, Mercy, "Sasha" Glen, Caleb and Jordan. They have been my motivation to work harder, explore more, and see it all through younger eyes. I would choose each one of you, every time.

And I am thankful for Lelya, my wife and my best friend. Whoever coined the phrase "better half" must have had a partner like her. Her strength and grace has kept our family and our business together. She holds the world together for so many people. And she holds my world together so I can go fly airplanes, chase bears around with my camera, and harass trout with a fly rod; and always come back knowing that the business and our family were left in hands more capable than my own. Mostly though, I just love the fact that no matter how big of an adventure I get to go on each day, I get to come home to her. And that tops all the other adventures, hands down.

Part I:
The Salvelinus, the Sockeye, and a Diverse Watershed

Prologue: How Stories Change Us

I admit that the stories of an angler may be less trustworthy than those of a criminal under interrogation. An angler walking along a favorite stretch of river after a great day of fly casting, if asked by a stranger how the fishing was, might respond with a shrug and a non-committal "nothing to speak of." Later that evening, however, the same angler might be heard boasting to friends of a great day on the water (though without any details of the exact location), perhaps embellishing the size of the catch while also exaggerating their own skill in making the day so successful.

But the *memories* of an angler (as opposed to their stories) could rival in accuracy the most objective journalistic reporting. What angler can't remember the vivid details of some particular fish they caught or failed to catch? The exact time of day. What insects were hatching. The water level. The fly they chose (with excruciating detail about *why* they chose that particular fly and not some other fly). And where the fish was rising (along the seam behind the boulder, just where the shadow of the big oak touched the water). They likely also know how much it rained the day before, where they stopped that morning for coffee and pastries, and the exact list (in order) of every other fly they desperately chucked at the fish that escaped and then stubbornly refused to eat.

My own memory takes me back to a mid-August morning in 2009. Despite the passing of time, the images remain fresh. Among other things, it's the day I made an important formal acquaintance with a small but beautiful fish known as a Dolly Varden. At the time of the memory, I knew little or nothing about this Alaskan species of char. I also knew little about Bristol Bay; I could not have named the national and state parks or wildlife refuges in its watershed. I hadn't yet heard of Pebble Mine.

A year earlier, my brother Ted had moved to Anchorage with his wife Susie and their college-age sons Brad and Michael. It was my first visit to see them since their move. Ted and I had just spent a night camping on a ridge between Symphony Lake and Eagle Lake in Chugach State Park. We'd caught several Arctic grayling in Symphony Lake and eaten wild blueberries, crowberries, and cranberries picked within a few yards of our tent. I'd also watched Ted rescue a ptarmigan entangled in monofilament line left on the shore by a careless angler. Realizing that the bird was trapped, he had carefully approached the terrified creature, held it as gently as he could with one hand while it struggled and attempted to peck him, and with his other hand cut the line free.

Now, in my memory of the following morning, we're casting flies again, this time in a long deep pool just a hundred yards downstream of Eagle Lake at the upper end of the South Fork of Eagle River. A vast brushy alpine meadow stretches for miles to the west as the valley slowly descends toward Cook Inlet. Though a few trees are scattered in sheltered hollows further down the valley, we are mostly above the timberline, far upriver of the reach of spawning salmon. To the north and south, the land rises from the river up the steep ridges that enclose the valley. Thick riparian brush gives way to wildflowers and low blueberry bushes, then to grass, lichen, and finally bare rock. To the east, jagged peaks of bare rock tower above us hiding glaciers and snowpacks, and forming an imposing wall around the lakes where we camped and fished the previous evening. As if the stark beauty of the surrounding landscape isn't enough to distract us from our pursuit of fish, a cow and calf moose appear and wade across the opposite end of the pool just two dozen yards away, oblivious to—or ambivalent about—our presence. We pause for several minutes to watch them.

When the two moose depart, our focus returns to the fish. The water, laden with glacial flour, is grayish turquoise with only about eighteen inches of visibility. A dozen years later, I will stand at the same pool on a warm mid-July morning when so many insects are hatching, and so many fish are feeding on those insects, that the water appears to be boiling. I'll entice some of those fish to take a little dry fly, and then invite my friend Josh to borrow my fly rod and catch his first ever Dolly Varden in this same pool. In the 2009 day I am remembering, however—my first visit to this spot, and the day of my own formal introduction to Dolly Varden and her cousins—autumn is already setting in. If there are insects hatching, we don't see them. The air is cold and damp, and any fish that are feeding are doing so below the surface. With low visibility in the silty water, I may have to bump a fish on the nose with my bead-head nymph to get its attention.

Eventually, however, I figure out the fly, the current, and the drift, and I begin to entice some strikes. Soon after, I have the privilege of meeting my first southern Dolly Varden: a beautiful ten-inch fish native to this little river. Though I haven't foreseen how this species of fish will change the course of my writing in the years to come, I nonetheless admire its beauty for a moment as I hold it in the water. I snap a quick photo, then open my fingers and let it swim away.

We catch a few more Dollies in this pool as we wait for Ted's son Michael to hike in and join us. Most of the fish are six or seven inches long, but we see an occasional ten-inch behemoth making use of the deep water and abundant insect life to grow larger than most of its relatives in the valley. When Michael arrives, the three of us spend the rest of the day fishing our way several miles downstream across the meadow as we backpack our way out to the car. We catch several more five-to-seven-inch Dollies. Over the next decade I will catch many more silvery hand-sized Dolly Varden in various turquoise-tinted flour-laden mountain streams in Chugach State Park, as well as some larger and darker southern Dollies on the Kenai Peninsula in October below spawning silver salmon and steelhead.

Poet, essayist, and novelist Wendell Berry, in his beautiful essay "A Native Hill", describes an experience of walking through the woods near his home and coming upon a patch of bluebells strewn about the forest floor. “In the cool sunlight and the lacy shadows of the spring woods the blueness of those flowers, their elegant shape, their delicate fresh scent kept me standing and looking. I found a rich delight in them that I cannot describe and that I will never forget.” He goes on to describe his feelings. “Though I had been familiar for years with most of the spring woods flowers, I had never seen these and had not known they grew here. Looking at them, I felt a strange loss and sorrow that I had never seen them before. But I was also exultant that I saw them now—that they were here.” He continues in the next paragraph:

> For me, in the thought of them will always be the sense of the joyful surprise with which I found them—the sense that came suddenly to me then that the world is blessed beyond my understanding, more abundantly than I will ever know. What lives are still ahead of me here to be discovered and exulted in, tomorrow, or in twenty years? What wonder will be found here on the morning after my death? Though as a man I inherit great evils and the possibility of great loss and suffering, I know that my life is blessed and graced by the yearly flowering of the bluebells. How perfect they are! In their presence I am humble and joyful.

Berry's description of his experience of those flowers resonated with me. The joy and delight, but also a sense of loss and sorrow. And perhaps more than anything, a deep longing.

In an essay for the journal *Written River*, I use the Welsh word *hiraeth* to describe my feelings about the stark majesty of that scene on the South Fork of the Eagle River, and the exquisite beauty of both the landscape and the fish I caught there. *Hiraeth* has no exact equivalent in modern English; the Welsh word suggests a powerful sense of wonder, longing, and hope, but also grief or sadness over something that has been lost. Perhaps the closest equivalent is the German word *sehnsucht*. In his book *The Weight of Glory*, twentieth-century author C.S.Lewis (best known for *The Lion, the Witch, and the Wardrobe* and his other children's fantasy stories set in the imaginary world of *Narnia*) defines *sehnsucht* as "the inconsolable longing in the heart for we know not what." The word or concept appears repeatedly in Lewis's writings, particularly when he describes his feelings after coming up a particularly beautiful place, scent, or sound. In *The Weight of Glory*, he describes such experiences as eliciting "the scent of a flower we have not found, the echo of a tune we have not heard, news from a country we have never visited." In *Pilgrim's Regress* he speaks of "that unnameable something, desire for which pierces us like a rapier at the smell of bonfire, the sound of wild ducks flying overhead, . . . the morning cobwebs in late summer, or the noise of falling waves." These experiences were, for C.S.Lewis, pointers to Heaven.

I have felt the same way.

Six years later, in 2015, I began intentional work[1] on this book with a trip to *The Farm Lodge* in Port Alsworth (Alaska) in the awe-inspiring landscapes of Lake Clark National Park and Preserve. The cold-water species of fish now commonly known as Dolly Varden char, or Dolly Varden trout, or simply Dolly Varden—a name taken from a character in the nineteenth century Charles Dickens novel *Barnaby Rudge* and first applied to another species of char known as bull trout—had become the focus of my writing. By then I've learned more about the Bristol Bay drainage, and about the proposed Pebble Mine. I've also learned quite a bit more about Dolly Varden char through books, scientific journals, and interviews with fisheries biologists and stream ecologists. Most poignantly and enjoyably, I've learned experientially from many more personal fly-fishing encounters with both Dolly Varden char and Bristol Bay in a wide variety of waters (and with a wide variety of flies).

And the more I've learned about the Dolly Varden species—about its natural history and diverse life histories,

1 Although I consciously began work on this book in 2015, I was unknowingly being prepared for it at least a dozen years earlier when I took my first wilderness camping and fly-fishing trip to the Bristol Bay region with my father. My thinking was also being shaped as I co-authored *Ents, Elves, and Eriador: the Environmental Vision of J.R.R. Tolkien* (2006) with Jonathan Evans, *Narnia and the Fields of Arbol: the Environmental Vision of C.S.Lewis* (2009) with David O'Hara, and *Downstream: Reflections on Brook Trout, Fly Fishing, and the Waters of Appalachia* (2014) also with David O'Hara. Some of my writings about rivers and fish in the Bristol Bay drainage—particularly those with less emphasis on fly fishing—appeared in *The Voices of Rivers: Reflections on Places Wild and Almost Wild* (2019).

and about the ecologies of the places it lives—the more this beautiful, adaptable fish has inspired in me a sense of delight, awe, and wonder with its magenta pearls, ruby red coat, and bright orange lipstick. The fish and the places it lives has repeatedly elicited in me that sense of *hiraeth*, or *sehnsucht*. As I've continued to learn more about the past and present of the species, I've also became curious about its prospects for the future in an increasingly developed world with a rapidly changing climate.

My thought going into this book, therefore, was that Dolly Varden char would be the narrative lens through which I could learn and write more broadly about the waters of the Bristol Bay drainage and the landscapes surrounding those waters. From small mountain creeks to powerful rivers, little tarns to some of the largest lakes in the world, crystal clear water to turquoise jewels laden with glacial flour, and from freshwater creeks far from the sea to brackish estuaries, the various waters that Dollies call home are really the most important characters of this story and much of the reason for this book. Yet the story is also partly my own: a story of learning more about these creatures and habitats through the work of scientists who have studied them professionally, the stories of people who have spent their lives around them, and through own exploration with a fly rod, a camera, and a willingness to be quiet, attentive, and present.

Stories, however, have a way of changing. Or perhaps it is more accurate to say that stories change their story tell-ers. To tell a story well one must enter that story. The more deeply a writer (and reader) imaginatively becomes a part of a story and allows it to change them, the better the writing (and the reading) can become. And entering a story, whether as a writer or a reader, requires a letting go of part of yourself; it requires a sort of identification with the characters. It is not an exaggeration to say that the story becomes you and you become the story. For writers, this is especially true of personal narrative. When the author is also a character in the story, the character development includes the author himself or herself. The author *ought* to change. This should also be true of place-based writing. The landscapes where we spend meaningful time have the power to change us. The more we inhabit those landscapes and the more present we are within them, the more we allow that.[2]

This book is both placed-based and personal. So it was that the story changed me as I wrote it, even as the story itself changed in the telling. What started as a book about Dolly Varden grew to be a book about three species of char native to the Bristol Bay drainage. But one can't really write about the waters of the Bristol Bay drainage without writing about salmon (especially sockeye). In many ways sockeye

2 That the stories we write should change us—most especially in narratives that are both place-based and personal—was brought home to me a decade ago by my wise friend John Elder as I was writing *Downstream: Reflection on Brook Trout, Fly Fishing, and the Waters of Appalachia.* A Vermont-based author of several beautiful placed-based works of personal narrative including *Reading the Mountains of Home* and *The Frog Run*, John asked me how my experiences and my writing of my own work-in-progress had changed me. It was great question, and it helped me articulate what had perhaps only been intuitive. Thinking through my answers also helped in the final stages of shaping that earlier book even as it does now with this book.

salmon are the most important fish of Bristol Bay, and of this book. They are a big part of the reason the stories told here are important. The lives of sockeye salmon are interwoven with the lives of the three species of char through the waters they co-inhabit for at least some of their lives. Rainbow trout (the member of the *Salmonidae* family which may be the most sought-after fish by fly-casters in Alaska) also find their stories interwoven with char and sockeye salmon, as do two other fish species that fly-fishers often chase: Arctic grayling and northern pike. They all are characters in this narrative, though some have fewer lines of dialogue.

1.
Alaska's Salvelinus at the End of a Line: A Note from an Author and Angler

This story begins with the genus *Salvelinus*—a member of the *Salmonidae* family known to me from my earliest days fishing when, as an eight-year-old, I joined my father on the Allagash Wilderness Waterway in northern Maine and encountered wild brook trout (*S. fontanilis*) in their native habitat. The genus includes the Dolly Varden, known by the scientific designation *S. malma.* Although many *Salvelinus* species are often referred to as trout, the genus is more accurately and commonly associated with the designation *char* or *charr.* The genus name comes from a German word meaning "little salmon."

A Primer on Char

In his classic text *Trout and Salmon of North America*, Robert Behnke notes that the name char is derived from a Celtic word for blood. Though he doesn't explicitly state it in his text, he was likely referring to the word *cera*, which is of Gaelic origin. Some guides to baby names, for example, give *cera* as one possible origin of the name *Ceara*, and give "bright red" as a meaning, though "blood red" might be a more accurate rendering of the earlier meaning. *Cera* and *ceara* are related to an older word for "spear". Although folk etymologies are notoriously unreliable, it is not difficult to imagine a connection between the meanings. Did the word for *spear* come also to describe the blood so often spilled by a spear in warfare, and then for the color of that blood? Surely whoever sat a few thousand years ago looking at the beautiful bright underbelly of a spawning male of some species of *Salvelinus* and named the fish *cera*—or as we would say today, *char*—must have seen in that fish a reminder of *blood.* Did that ancient Celt also think what Behnke thought, that the blood-red creature in front of him was one "of the most beautiful of all fishes"?

If the scene played out in the lands of one of the ancient Celtic tribes, the beautiful blood-red fish was mostly likely an Arctic char. But the description could certainly have fit any of several other species of char, especially brook trout or Dolly Varden. The species we now know as Dolly Varden likely evolved in Asia. They are believed to have come to North America following the land bridge some 40,000 years ago during the last ice age (in the same era as brown bears and a large human migration). Thus, among those species of *Salmonidae* (which includes all salmon, trout, and char) that we might describe as "native" to the North American continent—those that found their way here without human intervention—Dolly Varden were the last immigrants to arrive. Following the coast of the land bridge, Dollies made their home in the northwest corner of the continent, moving both directions along the coast while also migrating inland to colonize new waters. Puget Sound is as far south as they were able to reach. Coming late into the continent from the west, they never made it to New England like their char cousins Arctic char, brook trout, and lake trout did. Nor did they reach the Rocky Mountains as did bull trout (another North American species of char). As a result, northwest Washington is the only place outside Alaska that Dolly Varden can be found in the United States. In Canada, they appear only in the western part of the country: in western British Columbia and in drainages of the Beaufort Sea in the Yukon and Northwest Territories.

Alaska has two distinct subspecies of Dolly Varden, known (accurately, if not poetically) as northern Dolly Varden (*S. malma malma*) and southern Dolly Varden (*S. malma lordi*). They have probably been isolated from one other since shortly after they arrived on the continent. Though their natural and evolutionary histories are shared until relatively recently (in geologic time scales), there are numerous distinctions between them. They have separate mitochondrial DNA lineage, a different number of chromosomes (82 for *S.m.lordi* and only 78 for *S.m.malma*), and different ranges in numbers of vertebrae and gill rakers, with the larger northern Dollies having as many as eight more vertebrae and eight more gill rakers than southern Dollies.

Dolly Varden char are also diverse in another way: they can thrive in several different habitats and exhibit different life histories. *Fluvial* (river-dwelling) forms live in freshwater rivers and streams, and generally spawn in smaller streams. *Adfluvial* populations (particularly among southern Dollies) live much of the year in deep lakes, but spawn up tributary streams or move into rivers periodically to feed.

And there are also *anadromous* or semi-anadromous forms that migrate out to the ocean for parts of their life cycles. Even these sea-run forms might be further subdivided. Anadromous northern populations live much of their adult lives in the saltwater of the Chukchi Sea, moving back and forth over a thousand miles between Russia and Alaska and coming into fresh water only to spawn or overwinter and feed on salmon eggs. Some of these spawn in Russia and overwinter in Alaska. Some spawn in Alaska and overwinter in Russia. Some are summer spawners and some are fall spawners. And while these anadromous Dollies can grow to over twenty pounds on their marine diets of krill and their freshwater diets of salmon eggs, they grow slowly in the cold water and spawn only in alternate years. There are also southern populations of Dollies sometimes referred to as *semi-anadromous* that spend some time in saltwater, but (unlike their northern cousins) stay close to the coast and move in and out of estuaries.

Dolly Varden found in the Bristol Bay drainage—as with all waters that flow into the sea north of the Alaska Peninsula—are of the northern form. In addition to *S.malma malma,* two other species of *Salvelinus* are also native to the waters of Bristol Bay:[3] the circumpolar Arctic

3 There has been considerable historical confusion among three species of char: Dolly Varden, Arctic char, and bull trout. Some confusion has persisted to the present day, and many questions remain unanswered. Due in part to the diversity among different populations of char, especially Dolly Varden and Arctic char, there is passionate disagreement among scientists which populations should be considered different subspecies or even different species. (There isn't even agreement whether char should be spelled as char or charr.) Interested readers are directed to Behnke's excellent work for a good overview.

char (*S.alpinus*), which are the northernmost freshwater fish in the world, and lake trout (*S.namaycush*), which were the most widely distributed native species of trout or char in North America.

With lake trout widespread across all of Canada, native to northern New England and the entire Great Lakes basin, and introduced to many more states west of the Rockies where they are well-established (and in places having a devastating impact as an invasive species on native species such as cutthroat trout and bull trout), they are the least unique of the char species to a story about Bristol Bay. As result, they are left as minor characters. Nonetheless, they are worth mentioning. In their native range in the northeast United States (in the waters I grew up fishing for brook trout), they are known as *togue*. In the Great Lakes and westward, they are called *mackinaw* after an Algonquian place name meaning "many turtles". I have caught lake trout in lakes in Katmai National Park and Preserve where they were feeding below spawning sockeye salmon at the mouths of streams, and in lakes in Lake Clark National Park above the reach of spawning salmon where they co-exist with Arctic char. Although lake trout are a *lacustrine* species—spawning almost entirely in lakes—they will enter rivers to feed as well as to migrate and colonize new waters. They also have the lowest tolerance to salinity of any char, and many (including myself until recently) believed them to be one species of char that had no anadromous or semi-anadromous life histories. However, there have been occasional reports of lake trout in tidal water or brackish saltwater dating back more than a century. A recent study of four lakes in the Canadian Arctic—the "first detailed scientific data documenting anadromy in lake trout"—has found that many lake trout in these populations will make marine migrations starting on average at thirteen years of age. Despite the metabolic costs of adjusting between salinity levels, these anadromous lake trout were found to be healthier than the purely lake-dwelling residents of those same populations.[4] The mouth of the Naknek River in Bristol Bay is one of the places where lake trout have also been observed in brackish tidewater.

In contrast with the widely distributed lake trout, Alaska is the last major stronghold in the United States for both Arctic char and (as already noted) Dolly Varden char. While a subspecies of Arctic char known as blueback trout were once found in several northern New England lakes, they have been extirpated from almost all their native waters in the northeastern United States. In Canada, populations of Arctic char are focused in the north—true to their name, in and near the Arctic circle. The farther south one goes in Canada the more common lake trout become, along with brook trout (in the east) and bull trout (in the west) as well as both native and stocked rainbow trout and cutthroat trout (which are not char).

As much or more could be written about the Arctic char as about the Dolly Varden char, particularly with respect to diverse forms and life histories. Arctic char, like lake trout, tend to be lacustrine, spending most of their lives

---

4 For more information, see Swanson, et al, "Anadromy in Arctic populations of lake trout (Salvelinus namaycush)".

as lake-dwellers and typically spawning on deep shoals in the lakes. However, some Arctic char spawn in freshwater streams, and there are many anadromous populations that migrate out to sea beginning when they are five years old. Some lakes have two seemingly independent populations of Arctic char: a so-called dwarf form that rarely exceeds a foot in length and feeds on insects and snails, and a much larger predatory form that can grow to several pounds. I have fished in lakes in the Bristol Bay drainage with populations of both types of Arctic char, as well as lakes with populations of both lake trout and Arctic char. And while Dolly Varden char rarely thrive in lakes with Arctic char because the Arctic char outcompete them for food, they do occasionally overlap in territory, especially when Arctic char move into rivers to feed. There are some waters (including in Bristol Bay) where any of Dolly Varden char, Arctic char, or lake trout can be found at certain times.

Geography and Scope: The Bristol Bay Drainage

Although many of my early explorations for this book were motivated by the proposed Pebble Mine and began on Bristol Bay rivers closest to the footprint of that mine, the native range of the Dolly Varden species spans much of Alaska. And, as suggested above, their habitat includes a surprisingly wide variety of waters. Alaska is hopelessly large to be the geographic scope of a book. According to the

Alaska Department of Fish and Game, the state has more than 12,000 rivers and (astoundingly) more than three million lakes of at least five acres. Alaskans enjoy poking fun at Texans by pointing out that if Alaska were cut in half, then Texas would no longer be the second largest state; it would become only the third largest state behind the two halves of Alaska. So—although my research on Dolly Varden char took me to the Chugach National Forest and down along the Kenai Peninsula, to Chugach State Park and around the Anchorage area, up the Susitna drainage into Denali State Park, up the Matanuska River to its namesake glacier, and to some rivers and state parks in the Juneau area, and though my time in those places contributed to my learning—the focus of this project eventually narrowed back down to the Bristol Bay drainage where it started.

And that is still a hopelessly large area to cover. Bristol Bay includes most of Lake Clark National Park and Preserve as well as Katmai National Park and Preserve. Each of these alone covers more than four million acres. Their combined area could easily contain all off Connecticut, Delaware, and Rhode Island with plenty of room to spare. Although both parks extend east past the Alaska Range and into the Cook Inlet drainage, most of their waters flow into Bristol Bay. The Bristol Bay drainage also includes Wood-Tikchik State Park, which at 1.6 million acres is the largest state park in the country. And those three parks are just a portion of the area of the Bristol Bay drainage. The Togiak National Wildlife Refuge west of Wood-Tikchik State Park extends southeast into the Bristol Bay Drainage, and both the Becharof and Ugashik National Wildlife Refuges near the top of the Alaska Peninsula include rivers flowing westward into Bristol Bay. The drainage also includes abundant additional state land, private land, and tribal land. It is a vast landscape with so many rivers and lakes that many do not yet even have names that appear on a map.

All I really hoped for, therefore, was to provide some "snapshots" of the region: a few little colorful tiles, which if combined with enough other tiles might gave a mosaic portrait of the region. When you catch a glimpse of some wild and elusive forest creature, you don't need to have a complete scientific knowledge of its species—or even to see the whole thing clearly in a single sighting—to make your heart beat a little faster. You can get a sense of its size, power, speed and beauty while still leaving much of its life shrouded in mystery. Such a glimpse might make you want to see it more closely and learn about it. That's not a bad thing. Perhaps a more appropriate metaphor would be trying to get an idea of what a Dolly Varden char looks like by looking at a closeup painting of one of its magenta spots, and perhaps another painting of a little patch of its tail, and another closeup of the lips or of the kype of male. Even with a half-dozen close-up paintings, you'd still need to see a lot more to have a full picture of a Dolly Varden. But at least you might start to appreciate its beauty. My hope is that this glimpse of Bristol Bay will provide delight (both for those who have visited the area and those who have not), and also prompt in readers a desire to care about that place and its creatures—and its indigenous cultures, although describing the richness and diversity of those is far beyond the already overly ambitious scope of this book. I hope it will help you, as a reader, to understand the importance of protecting these places even if you never visit there yourself.

The following chapter, in addition to recounting some experiences casting flies in a Bristol Bay river, also describes a conversation I had with fisheries biologist Daniel Schindler about the diversity of waters in the Bristol Bay drainage and why that diversity is so important. That conversation, though short, ended up shaping much of this book. It also convinced me that this "mosaic" approach—even though giving just a glimpse of a small fraction of a vast and magnificent landscape—could be worthwhile: providing first-hand experiences of the tremendous diversity of waters in that drainage, and why the headwaters of Bristol Bay are so worth studying, admiring, and preserving.

2.
Spawning Sockeye: Abundance and Diversity in the Bristol Bay Drainage

The clock says it's late afternoon, but the mid-July sun is hours from setting. Light rain speckles the river, and mist shrouds the distant peaks. The guided fishing boats have departed for the day, motoring clients out to a nearby lake where float planes will pick them up and return them to the warmth and comfort of a wilderness lodge. The guides—at least some of the younger ones—will continue by boat to a campsite on the lakeshore where they will sleep in tents and be ready to repeat the routine the next day.

With the anglers gone and the river quiet again, the osprey and mergansers have returned. The osprey soars high overhead in search of fish. Now and then it spots something promising in the water below: a rainbow trout perhaps, or a grayling, or whitefish, or maybe one of the smaller sockeye salmon that have recently entered the river. It hovers awkwardly in place in its best imitation of a hummingbird. Far more often than not, after a few seconds of hovering it returns to soaring and searching. I watch anyway, hoping to see it dive: an exhilarating plunge of sixty feet or more straight down into the water. I want to see it come up with a fish in its talons, labor back into the air with a heavy load, and fly off to its nest to feed its young. And for a few seconds, my hopes for such a moment rise as the osprey hovers and then tucks its wings and plummets downward. Twenty feet above the surface, however, it pulls out of its dive and climbs slowly back to soaring level. Whatever it had seen has disappeared, or it has decided instinctively that its odds of success were too low for the risk and effort. A few minutes later, it disappears downriver toward different hunting grounds.

The flock of common mergansers with their tufted brown heads are less spectacular in their hunting, but still fun to watch. They move up and down the current, sometimes in tight little mobs and other times in long lines, taking turns disappearing beneath the surface and reappearing with barely a splash or ripple. The mergansers are wary. Any small movement on the shore spooks them, and they hydroplane out of sight, their webbed feet looking like the rotating wheels of a cartoon animal. They are also uncannily good at disappearing under water the instant before I hit the shutter release on my camera to take a closeup of one.

After dinner, the rain lets up. I've had a fruitful day of watching, listening, and writing about the river in front of me: about the rich and abundant life in, around, and intimately connected to it. Now it's time to experience some of that life more directly: to study the river's piscine inhabitants and their feeding habits up close with the aid of my fly rod. Thanks to emergency repair adhesive one of the young guides sleeping in a tent dropped off for me earlier in the day, I've patched a long tear in the knee of my waders. At 8 pm, I wader up and head out onto the water. The river is five hundred feet across at the shallow gravel bar where it forms at the outlet of the lake. Last evening, casting an egg-sucking leech in calf-deep water most of the way across that bar, I landed one of the largest rainbow trout of my life: a fish far too big and fat for my net. I also landed a fat and fresh sockeye salmon even bigger than the trout. Both fish had put up tremendous fights in the strong mid-river current. It had taken considerable time to land them with my five-weight rod. Tonight, I'm casting my eight-weight so I can bring large fish more quickly to net without tiring them out. Despite the rod change, though, I try the same fly and same strategy again, working my way slowly across the width of the river.

A brown bear shows up on the far bank looking for a salmon dinner. I'm barely halfway across when I see it—still a safe distance from where it's decided to hunt. Even if it spots a fish and leaps into the water, I don't expect it to go more than a few yards from shore. It is not threatening toward me, nor I to it. Nonetheless, my cautious and respectful side takes over. Though it took a fair bit of effort to wade out to where I am, and I'd like to get at least another hundred feet to where I caught the big rainbow trout the previous evening, I stop and watch the bear, admiring its strength and hunting proficiency while also hoping it will move on up the shore rather than come my way. I remain ready to retreat if it does enter the river. Just loud enough to make sure it knows I'm here, I call out "Hey, bear." It doesn't bother to look. I call again. It doesn't react. Yet a minute or two later, either the bear gets hold of my scent, or it looks up and sees me, or it just doesn't see any readily available food. It turns and disappears back into the woods. I wait a few more cautious minutes before I continue fishing my way forward.

Unlike the previous evening, my trip across the river does not yield a strike. Although I'd like to drift a fly through a channel close to the far bank, I'm feeling intimated by both the bear that only recently disappeared into the woods and by the last dozen yards of swifter deeper water I'd need to cross to reach that channel. A hundred feet from the far bank, I turn around and start fishing my way back. Ironically, it is almost within casting distance of where I started that something sucks in my fly. Suddenly my rod is bending and shaking like a sapling in a hurricane as another heavy, hard-fighting fish strips out a lot of line fast. With the eight-weight rod, though, I'm able to get the fish under control more quickly. Soon a hefty sockeye salmon, still silvery and ocean-fresh, is finning in the current at my right hip. It is far too big for my trout net. After an awkward struggle to get a hold of it, I release it by hand and watch it bolt off.

I head back to shore and exchange my eight-weight and egg-sucking leech for my five-weight and an elk-hair caddis to work the quieter water near the shore. Ranger Alison Eskelin has told me I'm in the buggiest part of this watershed. I believe her. It's one of the rare times I've been in the Bristol Bay drainage with salmon present and I've still been fly-fishing for trout with insect patterns rather than imitations of eggs or rotting salmon flesh. The caddis fly proves successful.

Of Sockeye Salmon and Caddisflies, Both in Great Numbers

It's mid-July. Sockeye salmon are moving upriver. If I watch closely for even half a minute, I'm likely to see at least one break the surface. Most of them are rolling like porpoises, just barely coming out of the water. Now and then, however, a more acrobatic salmon will leap a foot or two into the air, perhaps with an acrobatic flip. Some of the splashes of these high jumpers can be heard halfway across the river even over the rumble of the current. No matter how many times I've heard that sound over the past hour, it still makes me turn and look and gets my heart pumping a little faster.

So now I'm no longer watching and listening from the shore; I'm once again out in the river with a fly rod in hand casting for rainbow trout. As I wade, I look down *into* the water. Below the surface is not the occasional porpoising salmon, but a steady procession of sockeye cruising past my calves and knees. Some are already wearing bright red spawning colors. These are easy to spot even two feet down. But once my gaze attunes to the underwater light and focuses at the right depth, I begin to see many more, and not just the bright red ones. Many are still silvery green, closer to what they've looked like in the ocean for most of their adult lives. Although they are most numerous in the channels by the shores, the splashes give evidence of salmon moving upriver all the way across several hundred feet of current. If I can see so many in just a narrow band of water by my knees in just a few moments of watching, it's hard to fathom how many pass by in the span of a day across the whole width of river. And this is just a single tributary of one of several major rivers that flow into Bristol Bay.

In a few days, many of these salmon will start dropping their eggs into the gravel—as many as five thousand per female. Shortly after spawning, the salmon will die. If they do their final job well, most of the eggs will sink into the gravel where they are safe from predators. Three to five months later, a little alevin will hatch from each egg and spend another few weeks in the safety of the gravel before it depletes its food sack and must emerge as a young fry and compete for food. But some eggs get caught in the current and drift downstream, never managing to slip down into the gravel. That's why the trout and grayling—the fish I'm hoping to catch as I stand casting flies—will soon turn their attention away from insects to eggs. It's an abundant source of protein delivered fresh from the ocean to their doorsteps every year: protein that, unlike insects or smaller fish, does not try to escape. The eggs are as irresistible to a trout as a homemade cookie sitting on the kitchen counter is to me. It's known to anglers as the "egg drop" and it has a dramatic impact on how we fish for the trout.

Thanks to this tremendous supply of nutritious and easy-to-catch ocean-sourced protein, the trout in this river are numerous and fat. Although they sustain themselves through the spring and early summer on insects, most of their annual growth comes from consuming eggs during the brief season when salmon are spawning. So I appreciate the importance of the egg drop! At the moment, however, I'm thankful that sockeye salmon have *not* yet started dropping eggs. It means trout are still focused on insects, feeding both on the surface and below. In the afternoons and evenings, the air along the shoreline is thick with caddisflies bouncing like yo-yos from the bushes to the river. In the mornings I've been happy drifting little nymphs along the bottom. In the evenings I've switched to dry flies and the action has been just as fast or faster. I find it much more satisfying than casting egg imitations.

A blond sow brown bear wanders up the shoreline with two spring cubs. The word "adorable" is cliché, but it's really the best word to describe the little brown fur balls. I watch them for a while. It's hard not to. The salmon are not yet in the shallows on their spawning redds, so right now they are not easy prey for a bear. Nonetheless, this sow leaps into the water and in a heartbeat manages to come out with dinner for the family. The cubs join her tearing into the tasty fish flesh. In five days and nights spent on the river, I had six different visits from adult bears: two from a sow with spring cubs, one from a sow with second-year cubs, and three from lone males. I was able to watch some of them for several minutes. And in all those ursine visits, this was the only time I saw a bear catch a salmon. Not surprisingly, the bears have not yet put on their winter fat. Like the trout, they are waiting for the salmon to start spawning. Unlike the trout, however, they won't be content to eat individual eggs that drift lose in the current. The bears will snatch the entire fish, and if they get a female they gobble up all five thousand eggs at once. When the sockeye salmon start spawning in shallow areas and become more vulnerable to predators, bears will catch several a day and gorge themselves on eggs as they fatten up for winter.

I arrive at my favorite stretch of the river near where I'm staying. On my right, out toward the middle as I look downstream, a deep trough shelters several fish. I start casting there and soon land three fat rainbow trout on little nymphs: #16 Copper Johns and bead-head pheasant tails. I could wade more and fish the farther edge of the trough where I've landed some of the biggest 'bows of the past few days. But the river is busy today. There have been three or four boats per day over the past five days, most with one guide and two anglers. Today, eight boats are working the river, all within sight of me. One of them is waiting directly upriver from me about forty yards. I suspect the guide would like to bring his clients down where I am to fish this trough.

So I oblige him. Maybe it's the guide who brought me the wader repair and I owe him a favor. I don't know. The guide is in the water on the upriver side of the boat acting as a human anchor, so I can't see him. Either way, though, I don't want to be selfish. Time to follow the Golden Rule and treat others the way I'd like to be treated. I back out of the trough and open it up for the boat to drift downstream. I turn my attention to the left where there is a much shallower run only calf to thigh deep against a bushy undercut bank. A couple big boulders a dozen feet out from the bank provide nice structure and variety. It's the warmest day of my trip. No rain. Lots of sun. The caddisflies are hatching in thick clouds much earlier in the day than they have been. I spot several trout actively feeding on the surface close to the shore. One big pair of lips is sucking in flies just inches from an undercut bank beneath a bush in a little alcove. To get a good presentation, I'll need to drop a fly right into the alcove giving it twenty inches of drag-free drift before it floats under the branch and over the fish. On a previous evening with a stiffer breeze, I'd tried casting to that fish, but each time the wind carried my fly several feet upstream of where my tippet landed.

Today only a slight breeze ruffles the water. Conditions are nearly perfect. Over the past several evenings, big fish have stolen or mangled most of my caddis imitations. I tie on my last one that is a reasonable match in size and color to what is dropping off the bushes. My first cast falls two feet short of the target, and the next one just a foot short. Neither spook the fish. The third cast lands just where it needs to. A two-foot-long rainbow sips in the caddis. I lift my rod and the fish is on. For a few seconds I think I've overestimated the trout's size. The fish just sits in the current and lets me pull it five feet off its spot. Then it seems to realize it's on a line. Suddenly it turns, and with a exhilarating burst of speed it muscles downriver into the heavy current, taking me just short of my backing. I hadn't overestimated it at all. Several minutes pass before I can bring it back up the current to the net and release it.

Over the next hour I work seventy-five yards of the bank and catch another half dozen rainbows on dries, mostly within five feet of the shore. It's my last morning on this river. During my stay, I've landed more rainbows and grayling than I can count, plus one whitefish. I've also landed a half dozen fair-hooked sockeye salmon, mostly on my five-weight rod, and most of which were still fresh. The one species I also hoped to catch but have not seen is an Arctic char. A guide had told me the char had moved out of the river back down to the lake, but would return as soon as the eggs started dropping. I won't get to see one on this trip. It's time for me to pack and leave.

The Importance of Geologic and Genetic Diversity: A Conversation with Dr. Schindler

Two days later I sit with ecologist and fishery biologist Dr. Daniel Schindler in a small building further down the watershed from where I've spent the past six days writing, watching, studying, and occasionally wetting a line. The drainage has some of the best rainbow trout fishing in the world. I've experienced it. Now I'm learning more about the river ecology of this region and why its fisheries are so productive and important. Schindler is a great person to learn from. He is professor with the School of Aquatic and Fishery Sciences at the University of Washington. He is also a principal investigator with the Alaska Salmon

Program: a research program that began in the 1940s and has a tremendous amount of longitudinal data on salmon in the region from a variety of studies spanning decades. Although Bristol Bay's native species of char have been the primary lens through which I've been exploring and learning about Bristol Bay, it's impossible to make sense of the ecology of these waters—or the terrestrial ecologies of the landscape, or the cultures and economies that have grown around them—without understanding the vital role played by salmon, especially the abundant and widespread sockeye salmon and the tremendous amount of nutrients they transport every year from the ocean into those freshwater and terrestrial ecosystems. The whole inland ecosystem depends heavily on those fish from the sea.

And the dependency goes in both directions. The terrestrial and freshwater ecosystems of the Bristol Bay watersheds are vital to the salmon populations, and thus to the oceans where the salmon spend their adult lives and provide an important food source for many larger marine creatures, and as well as for humans. Without spawning habitat—which has been disappearing up and down the coast on both sides of the continent—wild salmon will cease to exist.

Central to my conversation with Schindler[5] is the topic of how sockeye salmon populations adapt to conditions and micro-habitats of their natal streams. Schindler explains that there are more than sixty such local adapta-

5 I later follow up on the conversation by watching some of Prof. Schindler's recorded lectures and also the beautiful video "Mosaic—the Salmon Wilderness of Bristol Bay, Alaska" written by Daniel Schindler.

tions: differences between one sockeye salmon and another depending on the particular places where they emerged from their eggs and will later return to spawn: the places to which they were headed as I watched them move past me over the past several days, as well as the destinations of millions of other sockeye salmon across thousands of Bristol Bay rivers and streams through the summer and fall months. Two spawning waters as close as a mile apart can vary significantly and require a different set of adaptations in the sockeye that spawn there. Those adaptations determine the genetic fitness of the fish *for that particular water*: its likelihood of surviving all the challenges not only of life in the ocean, but also of the journey at the end of its life, and of successfully spawning in those waters so that it might pass on those genes and corresponding adaptations to a new generation.

I am curious to see the whole list of adaptations. However, Schindler doesn't even get to number ten because the first several are so fascinating and I'm asking too many questions. Among the most intuitive to me are body size (including body fat), and the timing of when they spawn. While female sockeye salmon select mates for size and bright coloration—i.e. they mate with the largest, reddest male they can find—a male sockeye that is *too* tall can tip over in a shallow stream and never reach the female that might otherwise appreciate its size. Populations of salmon that spawn in the shallower streams have thus adapted over generations to have more streamlined bodies.

Just a few days after my conversation with Schindler, this point is brought home to me. I am standing beside a tiny brook at the top of the Kenai River system. The brook is narrow enough to leap over without getting my shoes wet, and only four to five inches deep. Upstream of me, the water comes riffling out from beneath a low ceiling of bushes. Two bright red sockeye salmon, one male and one female, are paired up just below the bushes as though waiting to shoot upstream into the shadows. The water is shallow enough that the backs of both stick well above the surface. I could easily reach out and slap one. Out of curiosity, I duck down and gaze up through a gap in the bushes. I count four more salmon under the protective cover of the branches, all bright red and half out of the water. None of the six I see have particularly tall bodies for sockeye. They look more like torpedoes than the sockeye salmon shapes I'm used to seeing.

I walk downriver forty yards and find a gravelly stretch where the stream is twice as wide and only an inch deep. A lone male salmon sits at the bottom of this riffle, apparently waiting to move upstream. This sockeye salmon is much taller than the other six. It has a high humped back. My approach spooks it, and it tries to swim upstream. It makes it only a few feet and then tips over in the shallow water: the same water that the other six salmon have already successfully navigated. It somehow manages, in a display both humorous and sad, to flop on its side another eight feet upriver before it loses momentum and flops back to where I first saw it, no closer to its goal than before all that effort. I don't know whether this salmon will mate. Maybe heavy rain will fall and raise the stream level allowing it upriver before all the females have mated, and it will get a chance to pass on its tall genes. But maybe not. If not, then the more of the torpedo-like genes and fewer of the tall ones will get passed on to the next generation that will return to this little creek.

On the other hand, sockeye salmon that spawn very long distances up swift currents do benefit from larger reserves of fat. The Kenai River salmon I watched had a journey of about 100 miles from ocean to the little creek where I saw them. Sockeye spawning up the Copper River may travel two to three times as far up more turbulent water to reach their spawning grounds. Studies have shown a correlation between fish size and migration success up the Copper River; bigger salmon are more likely to make it. (I've also been told that extra fat is why the famous Copper River salmon taste so good.)

Along with the importance of size, there is also a narrow temperature range of water where sockeye salmon eggs thrive and hatch, and where the emerging alevin also thrive. And the timing of when different rivers reach those ideal temperatures in late summer or fall can vary greatly, depending on whether the water flows down off glaciers or out of flat wetlands. The timing of when salmon spawn is thus vital. Even the size of sockeye salmon eggs varies based on their adaptations to a particular water. While larger eggs have some survival advantages, salmon spawning in streams with finer gravel need to lay smaller eggs that will fit down into the little gaps where they—and the young alevin that hatch from them—will be safer from predators.

And then we come to adaptations in MHC: the *major histocompatibility complex*, which are genes that play important roles in the adaptive immune system, including signaling infections to T-cells so the T-cells can respond. This was when our discussion leapt past my understanding of biology and genetics, but I nonetheless found it fascinating especially since human T-cells have been in the news since the start of the Covid-19 pandemic. MHC in fish has been the subject of much recent study including with respect to the impact of climate change. As the environment changes, so do the pathogens present in a water. The changing pathogens in turn require a shift in the immune systems of fish spawning in those waters. Different genetic adaptations in the MHC alleles in sockeye can have a significant impact on susceptibility to different infections (and resulting mortality from those infections), which can vary from water to water and season to season.[6]

Given the variety of waters and the number of adaptations necessary for salmon to thrive in different types of water, it is truly remarkable how successfully the sockeye species has colonized the Bristol Bay drainage. The Nushagak and Wood River system where Schindler does much of his work is particularly complex. But while some

6 For further reading, see Grimholt et al. "MHC class I evolution," and Dionne et al. "MHC standing genetic variation and pathogen resistance in wild Atlantic salmon."

river systems in the Bristol Bay drainage are more complex than others, to some degree that complexity plays out through the whole drainage including the two areas I have spent the most time: the Kvichak drainage which includes Lake Clark and its tributaries, and the rivers of Katmai National Park. That the Bristol Bay drainage has such tremendous landform and hydrologic diversity, and that sockeye are so adaptable, turns out to be important for several reasons. For one, it is vital to the broader ecosystem. Different conditions in different waters within the drainage mean that different places boom and bust in different years. Environmental conditions in a given year—a particularly wet season, for example, or an especially dry one; a hotter-than-average year or a cooler year—may make one stream worse for spawning, but because of the geologic diversity those same conditions will make another spawning area better that same year, thus stabilizing the overall spawning run across the region.

As I write this paragraph in 2022, I have recently learned that the run of salmon up the Kvichak River into Lake Iliamna was sizeable this year, but the number that continued from Lake Iliamna up the Newhalen River into Lake Clark was very low. The unusually wet summer and high water in Lake Clark might be one of the reasons. In late September, I took several walks to various points along the Tanalian River, a tributary of Lake Clark I have visited many times and written a couple articles about for fly fishing magazines. This late in the summer the river has usually come down considerably, but this year it remained unusually high. I spotted two sockeye salmon at the river mouth, apparently spawning on the gravelly lakeshore, as well as another pair a mile up the river pairing up in slower water near the bank. Four or five others appeared to be staging in a deep channel near the river mouth preparing to start upriver. Although only an anecdotal observation, this is far fewer than I have seen during the same season on a previous visit. I was also aware of how different the Tanalian River is from others I have been to in Bristol Bay, including a much later sockeye run than many rivers down in Katmai National Park and Preserve. I also had in my mind—largely because of my recent conversation with Schindler—the diverse spawning habitats I was observing even within the river. Most importantly, perhaps, is that while many locals were discussing the low return of sockeye salmon to Lake Clark, I was aware of the overall stability of the Kvichak River drainage and of Bristol Bay as a whole; many of those salmon entering Iliamna will spawn in other rivers and streams that might be *positively* impacted by the wet summer. (I was also aware that there were too few salmon in the river for me to bother casting egg-sucking leeches with reasonable hopes of attracting trout or char.) Thanks in large part to this protected diversity, sockeye salmon are thriving in Bristol Bay, even though salmon populations are suffering in many other watersheds. This is not only important to many other marine species that depend on salmon for food, but also to human food supplies. In 2021, Bristol Bay produced some seventy-one million sockeye salmon counting both the total harvested and the *escapement* (the number of fish left to spawn or feed the bears.) That was a record. But showing just how productive the area is, and how harvests are proving sustainable, the record lasted only a year. In 2022, the total was eighty-three million sockeye, of which sixty million (about 72%) were caught providing far and away the largest sockeye salmon harvest recorded.

The diversity in spawning habitats is also vital for another reason. Time is a limiting resource for predators such as bears, eagles, trout, char, and grayling. These predators binge on salmon and eggs, but can only do so only when salmon are actively spawning. Indeed, when salmon are spawning , they are so abundant that bears will largely ignore the flesh and strategically focus on the most protein-rich parts of the fish in order to maximize protein consumption in a limited window of time. Schindler noted that bears tend to eat only the brains of the male salmon, while consuming the eggs of the females. They simply discard the rest of the flesh: the lovely pink meat that humans are so fond of: grilled, smoked, or turned into lox. I've also witnessed birds—both gulls and eagles—targeting the eyes. What limits the total intake of protein for bears is not the amount available *per day* when the salmon are running, but how long the season is: how many *days* there are to feed. Having the spawning season spread out over multiple weeks and months is critical for predators. If all the sockeye spawned at the same time, predators would have only about three weeks to get all their fat storage for the entire winter.

Thanks to the diversity of spawning areas among various rivers, streams, and lakeshores in the Bristol Bay region, however, mobile predators have a much longer season. Bears move from waterfalls to streams to rivers to beaches following the timing of the various sockeye salmon runs. Eagles do the same thing. Even trout, Arctic grayling, Dolly Varden and Arctic char follow spawning salmon from place to place, swimming dozens to hundreds of miles around the system to follow the egg drop. Anglers know that some rivers with excellent rainbow trout and Dolly Varden fishing in August have terrible fishing the rest of the year because there are very few fish at other times; the trout and char move in to chase the salmon, and move out when the salmon are gone.

In the Wood River system, with all its diverse habitat protected by Wood-Tikchik State Park, the sockeye salmon spawning seasons is spread out two and a half to three months. Wood-Tikchik State Park on the Nushagak and Wood River watersheds is one of the best and most important examples of the diversity of habitats, with the Wood River drainage especially important. Since the year 2000, Wood River has been producing about eight million sockeye a year, with the Nushagak producing another four million. Two thirds of these end up harvested. Slope is the factor that impacts almost everything else including water temperature and gravel size. The drainage has a variety of rivers with steeper slopes, and others draining relatively flat watersheds, and many in between. The rivers on steeper slopes don't carry as much carbon, but they have lots of alder and are nitrogen rich and crystal clear. The flatter waters carry more carbon but are nitrogen pour and tea colored.[7] And the higher gradient rivers are often just a few miles from the flat ones—a fact that becomes evident when you examine a terrain image or topo map of the half dozen big lakes in the Wood River drainage, and compare their eastern and western ends.

The variety of spawning habitats also extends beyond just the differences in river gradients. There is spawning habitat on lakeshores and in larger rivers, as well as in the myriad small rivers and streams including cold streams draining mountains and warmer streams draining flat wetlands. These diverse spawning habitats and the different adaptations needed by each is what has led to the important genetic diversity among the millions of sockeye salmon that spawn up those waters. Schindler made the important point that this diversity plays out at small scales. That two streams only a mile apart can vary greatly was an oft-repeated refrain, that proved to be important for conservation reasons. Developers and the resource extraction

---

7 For many years the prevailing scientific belief, supported by chemical analysis of nitrogen isotopes, was that marine-derived nutrients (MDN)—nutrients carried upriver from the ocean by spawning salmon and deposited across the landscape in salmon carcasses and in the excrement and urine of bears (and other salmon consumers)— accounted for the majority of nitrogen available to riparian plants. While anadromous salmon do supply large amounts of nutrients and organic matter to inland ecosystems (both aquatic and riparian) including significant localized deposits of different forms of nitrogen, a variety of studies over the past two decades have provided evidence that the role of salmon-borne nitrogen actually available to plants has probably been exaggerated. In terms of nitrogen available to other riparian plants over the long term, alders are more likely the primary contributors.

industry have sometimes argued for project approval by claiming that the proposed development would protect a big river, and impact only myriad small streams. But the myriad small streams provide much of the diversity and overall stability that makes the Bristol Bay salmon fishery so abundant and resilient. As Schindler phrases it, *lots of little places matter a lot.*

Unlike much of the Pacific northwest in the lower forty-eight states, and even many parts of southcentral and southeast Alaska, Bristol Bay still has a largely intact watershed which has been protected from development and resource extraction. As noted earlier, Lake Clark National Park and Preserve, Katmai National Park and Preserve, and Wood-Tikchik State Park all sit entirely or largely on the Bristol Bay drainage, as do portions of three different wildlife refuges. These parks, preserves, and refuges protect a variety of diverse spawning habitats, making for one of the last great salmon strongholds in the world: a vast watershed that sockeye salmon have successfully colonized, with hydrologic and landform diversity resulting in resilience year to year and decade to decade, warm periods and cold periods, wet periods and dry ones.

And while I come to those places with a fly rod in hand, and delight in the incomparably grand majesty of the place as well as the beauty of the fat, spawning, male Dolly Varden char or egg-gorged football-shaped rainbow trout that will strike my flies in those waters, it is the importance of those diverse watersheds to the world that my thoughts often return to.

3.
First Encounters with Bristol Bay, Alaska's Salvelinus, and the Egg-Sucking Leech

Flashback. It is mid-July, on the third day of what is supposed to be a six-day float trip down a remote wilderness river in Alaska. I am still trying to catch an Arctic char on a fly rod. With me are my father and our guide Brad Roche who (at the time of this trip) runs a wilderness lodge out of Pedro Bay on Lake Iliamna. Two days ago, our float plane pilot landed his de Havilland Beaver in one of the kettle ponds that dot the landscape—the only one near our desired destination large enough for takeoffs and landings—and dropped the three of us off. Ducks swam out of the way as we taxied to the shore and unloaded all our gear: a steel bear-resistant box with a six-day supply of food, an inflatable three-person raft, and all our fishing and camping supplies and clothing. Then the pilot flew off with a promise to pick us up at the other end of the float in six days. Or at least he would if the weather permitted (which, as it turned out, it did not.) Otherwise, he would return at some unspecified time after that, whenever he was able to (which, thankfully, he did.) We then hauled our impressive amount of gear across the tundra landscape to the river, 1.25 miles away.

The landscape is flat. It is not quite pancake flat, or what east coasters think of as midwestern cornfield flat.

Although anything I might call a mountain is many miles to the west, a few low hills dot the terrain. Here and there the river curves around one of those hills, carving low bluffs which rise twenty feet above the riverbank. They are low enough to climb in half a minute, but high enough that from the top one can see for a long distance. Because everything beyond that bluff is flat. Windy tundra flat. In the rare moments when the river passes one of those bluffs, or drops a few feet below the level of the tundra where trees can grow in the lee of the prevailing wind, we find a few hardwoods, mostly alder. At one point near the end of the trip, we float into hillier terrain and we pass through a low-lying forest dominated by white spruce. Mostly, though, a wide swath of thick waist-high brush lines the riverbank, with grass and smaller sparser bushes spreading out beyond that as far as one can see. The wind and the cold make life nearly impossible for anything to grow more than a few feet off the ground.

But flat is still bumpy. The terrain is textured with foot-and-a-half tall clumps of grass. All the way from the kettle pond to the river I alternated: step up, step down, step up, step down. It seemed to double the actual distance. It took five trips across that 1.25 miles of tundra from pond to river heavily loaded with gear plus four empty return crossings from river back to the pond. Finally, we had all our gear at the riverbank. Brad inflated the raft, we loaded it, and started downriver.

Two days later I have become thankful for the constant breeze. Even though it makes casting flies difficult at times, it keeps the mosquitos off me. Thanks to the breeze, there is nary an insect to be seen or heard while we're floating the river, or walking along the gravel bars near our camps, or when I stop and cast flies (which I do as often as possible). But when I approach within a few feet of the shoreline brush, the collective buzzing of astronomical numbers of mosquitos is so loud that the first time I hear it I look up into the sky for an airplane. I've heard an Alaskan moose can lose a point of blood a day to mosquitos. Though that may be sustainable for a moose weighing eight hundred to sixteen hundred pounds, it is not sustainable for a two-hundred pound human. If any of us venture into the brush, we take our lives into our hands. Fortunately, I don't have to walk into the brush often, because I am trying to catch an Arctic char on a fly, and they stay in the river which is lined almost everywhere by a gravelly shoreline. Last night we camped on a gravelly island in the middle of two braided channels. After dinner and setting up camp, I had hundreds of yards to fish along the mosquito-free gravel upriver, downriver, and on both sides, as late as I wanted into the evening because the sky never grew truly dark.

Foul Hooked

For the past couple days I have been fly-fishing mostly for salmon, using an 8-wt rod. Three species of salmon are spawning in the river: Chinook salmon (which are also known as king salmon), chum salmon (which have a whole host of names including dog salmon, keta, and silverbrite), and sockeye salmon (which many Alaskans refer to simply as red salmon or "reds" for short). I have landed one Chinook weighing some forty pounds and several others ranging from ten to thirty pounds. The forty-pounder took me deep into my backing on its initial run, and I spent well over an hour fighting it before I finally landed it. When I hooked it, I was far enough upriver from our tent to be out of sight. By the time I landed it, I was out of sight of our tent around a bend on the downstream side. The fish dragged me a quarter mile or more downriver.

I have also caught more chum salmon than I can count. No salmon makes for good eating when they are far into their spawning run with their flesh starting to rot. Some Alaskans consider chum salmon undesirable for food even early in their spawning runs. In northern Alaska, however, where they are called silverbrite and keta, they are prized eating. There is also a good market for them in Japan. Regardless of their desirability for food, they are the second largest of the five species of Alaskan salmon, behind only the Chinook. And, unlike most sockeye, they keep predatory instincts well into their spawning runs. They will strike a big gaudy fly aggressively. This makes chum salmon a very enjoyable game fish. Although they sport big toothy grins that remind me of cartoon caricatures of junkyard dogs, and that can cut your fingers up when you retrieve a fly from their mouths, they are nonetheless a beautiful colorful fish. Unlike sockeye, Chinook, and silver salmon which often turn almost uniformly red, chum salmon have a beautiful coloration with vertical stripes of mottled greens and purples. They are also—like the other salmon they share the water with—an important part of this river's ecosystem, contributing marine-derived nutrients (MDN) in the form of eggs and dead carcasses. Their big bodies carry a lot of those nutrients. The first several I landed were a thrill.

After some time, however, I started avoiding the chum salmon. Or at least I *try* to avoid them. That isn't always easy. Sometimes I spook a school of chum and they turn as a group and swim right through my fly line. If I'm not able to get my fly out of the water quickly enough, then my line slides along the backs of one spooked fish after another until the whole school has passed. As often as not, the hook catches on a dorsal fin of one of those fish. Battling a fair-hooked fifteen-pound chum salmon that takes a fly in the mouth is an enjoyable challenge. A big salmon accidentally snagged somewhere *other* than in the mouth, however, can be extremely difficult if not impossible to turn around. I can easily lose fifteen to twenty minutes tiring a fin-hooked chum in hopes of eventually retrieving my fly, often without success. These fish are capable of swimming hundreds of miles against strong river currents to spawn; twenty minutes of swimming against the pressure of my fly rod does not tire them. Indeed, they seem barely to notice my efforts to turn them around. I often give up and intentionally break my line, sacrificing the fly. So I begin targeting the king salmon which are not moving in big schools and are not as prone to the accidental foul-hooking.

Atlantic salmon, along with Pacific steelhead (which are an anadromous strain of rainbow trout), do not necessarily die when they spawn. They can drop back down to the ocean and return to spawn again the following year, and sometimes even a third time, getting larger with each additional year in the ocean. Steelhead spawning seasons often overlap with salmon spawning in the Pacific northwest, and steelhead will sometimes stay in rivers for prolonged periods up to the entire winter—much longer than needed just for spawning—to take advantage of the abundant supply of salmon eggs. This is not merely a reminder of how many creatures depend upon, have adapted to, or at least greatly benefit from, salmon eggs as an important part of their diet. It is also to say that continuing to eat even while in freshwater is natural and important for the ocean-run steelhead and Atlantic salmon.

By contrast, all five species of Pacific salmon die when they spawn. When they leave the ocean and enter freshwater, they have fat reserves to carry them through the remainder of their lives. They don't *need* to eat any more. That's what my guide and many others have told me, anyway. But even if salmon are no longer eating for nutritional needs—even if all their remaining energy is being poured into spawning and not digestion—they still seem to have a feeding instinct built up over a lifetime. This would explain why even a hundred miles up some freshwater river, the salmon will strike a big ugly bright pink fly that looks nothing like any food available in that river, but which instead imitates some meal they would have found in the ocean weeks earlier. Silver, chum, and pink salmon especially will still chase flies aggressively even after they have moved into rivers on their spawning migrations. And while the Chinook and sockeye salmon I have caught on this trip might have only chased my flies out of aggression while defending their spawning redds, in years to come on a different river I will even catch sockeye on nymphs.

But this is my first time fly-fishing for salmon. I haven't learned any of that. I only know what our guide Brad tells me. That includes the fact that sockeye salmon are notoriously difficult to catch on a fly once they are in freshwater, and that many people either snag them or (in places where snagging is illegal) they use a technique called flossing: they manage to align their fishing line with the mouth of the salmon and the pull it through like floss so that the hook catches the fish in the mouth and is thus legal even though the fish had no intent of biting the fly. I'm not interested in flossing.

I'm interested in getting fish to chase flies. I somehow managed to catch a fresh sockeye salmon—fair hooked—on the first day of the trip. It was at least six pounds of good salmon meat. That was enough to provide dinner for the three of us for the rest of the trip. It is the only fish I will keep on the adventure. So I'm no longer casting for sockeye. The only salmon I am now targeting are the Chinooks—the king salmon. I'm looking for the really big ones. As noted, I have already caught several.

I suppose this is appropriate. I am on a river called the King Salmon River. It is a tributary of the Nushagak which starts up by the northeast corner of Wood-Tikchik State Park and flows eastward out of the foothills and across a wide flat plane into the Nushagak, from whence it's waters then flow south down into Bristol Bay at Dillingham. Even after I learn the name of the river, and its general location, it will be a couple years before I pinpoint on a map exactly where our trip took place. Brad has been secretive and asks me at the time not to mention the river in any magazines. He is exploring it for the first time and is pleased to verify that it hasn't yet gotten the pressure of so many other rivers in the region. On our six-day float, we will see only four other humans.

Twenty years later when I next return to the western tributaries of the Nushagak, things will have changed. Lakes are warmer than they were just a few decades earlier, with more zooplankton. Alders have been able to colonize more territory in places, but there have also been caterpillar outbreaks killing off large areas of subalpine alders. As with many areas in the Arctic, the tundra has become shrubbier. While sockeye salmon populations have been thriving throughout the region, Chinook salmon have seen a dramatic decline. Nobody seems quite sure why. Somebody reading this book may see in print the river name I left unmentioned for several years, and could find it on a map by following the Nushagak River upstream from Dillingham. But if that reader arranged a trip expecting the good king salmon fishing I had with my father twenty years ago, and expected to have the river to themselves on that trip, they would almost certainly be disappointed in both regards. Caribou populations have declined even more dramatically. In 2003 we spent over two minutes flying over a single long herd of caribou. Based on the speed at which we were flying, we estimated that the herd stretched out some three or more miles. Now caribou, like king salmon, have become rare in the Nushagak area—even in the King Salmon River. Some think the herd got too big and overate their supply of lichen. That's one theory. But climate change may also be playing a role. Willow trees are colonizing the area, producing great habitat for moose but doing nothing for the caribou.

At the time, though, I am not thinking about any of that. I have not yet come to understand the ecological importance of native fish species in general, or of salmon in Alaska in particular. I know nothing of sockeye salmon and their varieties of spawning habits. I haven't yet heard of Dolly Varden char, and I wouldn't know the difference between a Dolly and an Arctic char. The seeds for all of these interests are being planted on this trip. The landscape is speaking to me, beginning to instill in me an interest and a concern. But at present I am simply enjoying what seems like a once-in-a-lifetime experience of a wild, vast, beautiful landscape and some uncrowded fishing for a variety of fish species that I have never caught before—some of which are very large, and quite thrilling to catch. It is a trip my father and I spent three years planning and saving money for. Float planes in Alaska are not cheap. But I wanted to take this trip while my father was still able to. He is almost seventy years old at the time. And once on the trip, I take advantage of the long hours of daylight in Alaska, sometimes casting flies until almost midnight. I get four or five hookups per day on the river's namesake fish. I also land a few sockeye salmon, dozens of chum salmon, and so many Arctic grayling that Brad starts calling me the Grayling King.

When I don't see any Chinook salmon and am tired of landing chum salmon, I get out my 5-wt rod and try to catch one of the resident rainbow trout or char. It is likely that before this trip I have never caught a rainbow trout in its native habitat, though I have caught plenty of non-native rainbows in the Rockies and the northeast. I am certain that I have never caught an Arctic char, though I will later learn of their historic presence in some New England lakes.

The Egg-Sucking Leech

Some flies, like the *royal Wulff*, are named after a fly-tier who first created the pattern. Some have more whimsical names like *royal coachman*, *Mickey Finn*, *stimulator*, or *Dolly Llama*. Some are named after their ingredients, like the *pheasant-tail nymph*. Some are named after a mix of their ingredients and the thing they are trying to imitate, like an *elk-hair caddis*. I am casting a fly called, simply, a *blood-sucking leech*. That pretty much describes what the fly is trying to imitate.

In any fresh water in Alaska reachable by spawning salmon, the basic food web revolves around salmon eggs. Bears eat them. Birds eat them. Other fish eat them. Leeches eat them. So for a fly-fisher, egg patterns are a must in Alaska. In many waters, such as the famous Kenai River, many anglers simply thread an egg-colored egg-sized bead onto their line above a hook, which is often both cheaper and more effective than a fly tied to look like an egg. Fishing stores carry whole isles of beads of different sizes, colors, and patterns because different salmon, in different waters, have different sizes and colors of eggs. It's part of their adaptations. Even two sockeye salmon can have different shades and sizes of eggs depending on what water they are spawning in. Trout and char see a whole lot of eggs and can be very selective. In many situations, the egg-sucking leach can be even a step better than an imitation egg. "Two meals for the price of one," Brad tells me when he suggests the pattern. "Fish think they are getting an egg *and* a leech all in a single bite. Double the protein. Grayling, trout, and char can't resist them." This is my introduction to the pattern. A decade later when I start returning regularly to the Bristol Bay drainage, variations of egg-sucking leeches will be my staple fly. I develop several different patterns and variations of my own and tie a whole variety of them. Right now, though, I'm using one I bought in a fly shop. And I'm trying to catch an Arctic char.

Like my first few chum salmon, the first few grayling I catch prove very enjoyable. Though lacking in the flamboyant colors of the char genus, or even of a rainbow trout or spawning salmon, they have a subtle beauty: a shimmery incandescence, like an old-school jazz singer in a sequined olive gown. What might first appear as a uniform olive is really a mosaic of numerous shades of olive, silver, gold, and even blue, with a just a few scales of black. Their ventral fins, though easy to miss on a quick in-water release, are especially gorgeous; in a full-grown adult fish, they appear a nearly translucent yellow-green with soft purple stripes. What really catches the eye is their unique dorsal fin. Dark green and often fringed in red, it looks in shape and relative size more like a rounded-off sailfish fin than the fin of any other freshwater fish I have seen.

However, I can barely get an egg-sucking leach in a promising patch of water in pursuit of a trout or char without hooking a grayling. With the grayling so aggressive, I have little chance of catching my sought-after Arctic char. So after landing and releasing half a dozen grayling, I set my 5-wt back down and return to my salmon rod. Or I take a break from fishing altogether. This pattern repeats daily. In the end, I catch only half a dozen char on the entire ten-day trip to the Bristol Bay drainage. Three of these come on a short trip up the Iliamna River after our King Salmon float trip ends. And these three, as I will learn many years later—despite what Brad has told me at the time—are likely northern Dolly Varden char and not Arctic char, though at the time I have not yet learned about either species of Salvelinus.

The Iliamna River is full of sockeye salmon as we jetboat our way up, two days after the end of our float trip. The schools are far thicker than what we had witnessed on King Salmon River a few days earlier. They are piled up in the deep slow water at the river bends. In places they are so numerous they look like just a thick red smudge in the water. You can't even see the river bottom. From the air, they look like a red oil spill, or like puddles of blood. But these sockeye salmon in the Iliamna are not yet dropping eggs. They haven't reached their spawning redds, which is why they are still schooled up. And since they aren't dropping eggs, the egg-sucking leech proves ineffective here. When I spot a pair of char feeding on the inside of a bend, I watch them for a while. They are clearly feeding on insects drifting along the river bottom. I change my tactics altogether and switch my thinking from an Alaskan salmon stream to a little east coast brook trout stream. Removing the egg-sucking leech, I tie on an imitation caddis nymph which I dead-drift along the bottom. Sight-fishing in this way, I land my first two char of the trip. A few bends upriver, in a swirling eddy at the edge of a pool too deep and turbulent for my nymph tactic, I switch to a streamer fly. After several more hits and misses, I land a third char.

The Beautiful but Tragic Life of a North American Immigrant

As noted, at the time of this trip I have too little experience with Alaska's species of char to know whether they are *S.alpinus* (Arctic char) or *S.malma* (Dolly Varden char). More than a decade later, my love of the char genus nurtured in my youth fishing in Maine and then up and down the Appalachians for a species of char known as brook trout, will lead me back to Alaska far more interested in Dolly Varden char than I am even in Chinook salmon. That is also when I begin to learn the differences between the various char, their evolutionary histories, and even some of the human history of how scientist began to distinguish between them in Alaska.

Although I now have more confidence I can distinguish between Arctic char and Dolly Varden, they can at times be difficult to tell apart even for experts. And while they don't often overlap in territory—in part because Arctic char outcompete Dolly Varden, and so Dollies don't do well in the same waters as Arctic char—there are places where both can be found occupying different niches. Brad identifies the three fish I caught as Arctic char. A decade later, a fisheries biologist with Alaska Fish and Wildlife will hear where I caught them, look at my photographs, and tell me they are Dolly Varden. Since I am in a drainage north of the Alaskan Peninsula, this would make then northern Dollies. The confusion will eventually lead me to talk with several other char biologists in my effort to learn more about the differences in these two char species, and about their ecological and cultural importance in Alaskan rivers.

One of those biologists was Jack Dean whom I would meet in 2015. By that time, Dean already had several decades of experience researching species of char in southeastern Alaska. His last many years of work took place mostly on the Kenai Peninsula, but his research was world class, and he had knowledge about the char genus from around the world including Maine and Asia. A Korean War veteran, Dean described himself as an independent researcher. He'd had a 30-year career as fisheries biologist with U.S. Fish and Wildlife Service, with one of his accomplishments being his successful lobbying for catch-and-release fishing practices in Yellowstone National Park (where he served for seven years). However, he eventually shifted to independent unfunded research in order to be free of political purse strings. It seemed that some people didn't want to hear (or accept) his findings on Arctic char because they contracted popular opinions, and also had political ramifications including to hydro-electric projects (such as the one on Cooper Lake) which altered water levels of spawning habitat.[8] While every other lake in the area held Dolly Varden, Dean correctly identified the char in Cooper Lake as Arctic char rather than Dolly Varden. Arctic char are much rarer on the Kenai Peninsula, which is near the southern end of their range.

One of the things I liked about Jack Dean when we talked was that his method of research often involved making visits to lakes and sampling the fish populations with his fishing rod. Or in cases engaging Boy Scouts to collect some sample fish for him with similar methods. Here was somebody after my own heart. In some ways, I suppose he gave me the model for this book—not that I *harvested* fish with my fly rod, but only that I used it to have some close-up experiences with the fish I was learning about. Although Dean certainly took part in what are acknowledged as scientific surveys, much of his knowledge came from first-hand experience on those trips he often took with his wife in

8 This was Jack Dean's perspective on the controversy surrounding his findings. His story reminded me of when I was working with David O'Hara on our book *Downstream,* and we learned that the state of Kentucky officially denies that brook trout are native to the state. While there is some non-definitive evidence to the contrary—i.e. evidence of native brook trout pre-existing European settlement—it is possible that the claim is accurate. What seemed politically more important than scientific or historical evidence, however, was that the existence of native brook trout would have been a hindrance to the very lucrative practice of mountaintop removal for the coal industry.

their canoe. In addition to delighting in the stories of these experiences, I also came away from our conversations—and from some research papers he pointed me to—with quite a bit of new knowledge.

Although it was DNA analysis that verified his identification of fish in Cooper Lake as Arctic char, Jack described some of the visible and invisible differences between *S.alpinus* (Arctic char), *S.malma malma* (northern form Dolly Varden) and *S.malma lordi* (southern Dolly Varden) that had enabled him to distinguish between them in cases where it might have been ambiguous. Arctic char, he told me, almost always have a few spots larger than the pupil of its eye, while Dollies have smaller spots. He also said to check out the *caudal peduncle:* the narrow part of the body just above the tail. It is thicker on Dolly Varden (and also on brook trout), and narrower on an Arctic char. And Arctic char have a deeper fork in their tales (more like the char known as lake trout), while in Dollies and brook trout the tail is more square—which, indeed, leads to one of the famous nicknames of the brook trout: "squaretail". Unfortunately, the most definitive way of identification (other than genetic tests) requires killing the fish: Dollies have a fragile clear air bladder, while Arctic char have pale pink air bladders.

Dean also first pointed out to me what I would later read in several other sources (many of which likely benefited from Dean's work), that northern Dolly Varden and Arctic char both have 78 chromosomes, while southern Dollies have 82, and that Dolly Varden (both northern and southern) have more vertebrae than Arctic char. And he noted that they could also be generally (though not definitely) distinguished by differences in their gill rakers and *pyloric ceca*. What was more interesting to me—particularly since chromosomes, vertebrae, gill rakers, and pyloric ceca are not helpful to the normal observer wanting to identify a species without harming the fish—were the differences in behaviors and life histories and even between sub-populations of the same species in the same water. "Studies show that Artic char will naturally separate into size categories and utilize different parts of the lake," he told me. "Smaller ones struggle to survive eating mostly insects. The bigger ones have learned to eat sculpins and snails." After a pause, he adds, "This is an extremely plastic species."

Part of the plasticity of both Arctic char and Dolly Varden is in their spawning. Arctic char are generally more like lake trout than brook trout in that they are lacustrine in life history: they are lake spawners. In deep mountain lakes, they typically spawn seventy feet down in the rubble of a new landslide. Some females, Dean explains, will get wear on their fins from the gravel, but others show no wear at all and are likely just broadcasting their eggs rather than preparing a redd. But other Arctic char are stream spawners, which is part of what led to Dean's work trying to determine whether water levels from a dam were negatively impacting spawning. In one case, he found both large and dwarf Arctic char spawning in the same place, immediately in front of the outlet of a dam control structure.

Dean also noted that Arctic char feed on sticklebacks.

The threespine stickleback native to Alaska is a small fish—two to four inches in length—that has freshwater, marine, and anadromous populations. Freshwater sticklebacks feed on fish eggs and fry, invertebrates, and (in open water) plankton. This is of interest because loons carry a parasite that gets into char water and onto the plankton, from the plankton to the sticklebacks, and from sticklebacks into the char. But primarily into Arctic char. "Dolly Varden have very few parasites," Dean says. "They are probably not eating sticklebacks. Or if they get them, the freshwater parasites die when the Dollies go out to sea."

Although I would later read it in several other places, including in Behnke's book, it was also from Jack Dean that I first learned the tragic story of the state-sanctioned attempt at extirpating Dollies. Like many other populations of immigrants to the North American continent, Dolly Varden char were at one time unfairly blamed for societal woes. In particularly, they were blamed for hurting salmon populations by feeding on salmon fry. So in the 1920s and 1930s, the state of Alaska put a bounty on Dolly Varden, offering $0.02 to $0.05 each. By some estimates, six million Dolly Varden were killed and their tails turned in for bounty.

It is a sad story, with many levels of irony. Among other things, Dolly Varden char had coexisted with salmon for thousands of years. They do eat salmon eggs—a fact I have often taken advantage often with my egg-sucking leech flies—but they eat the eggs that are drifting loose and would not have hatched anyway. They also eat the flesh of dead salmon. Which is to say, they are good scavengers. However, Dollies are not very good at preying on salmon fry. According to the Alaska Department of Fish and Game, the one salmon species they seemed to have some success preying on are juvenile pink (humpy) salmon. In times and places when pink salmon fry are migrating out to sea in great abundance, Dolly Varden will successfully prey on them. But, as noted, they had been doing that for centuries and salmon were still successfully spawning. Arctic char and silver (coho) salmon turned out to be much more successful at preying on fry of other salmon such as sockeye. However, the more recent immigrants got the blame, and suffered the wrath.

In another level of irony, however a study of some 20,000 bounty tails indicated that more than half were from coho salmon. Others were from rainbow trout. So the attempt to *save* salmon by killing Dolly Varden, resulted in the slaughter of salmon—perhaps more than three million, if the statistics from those 20,000 tails extrapolate to all the six million bounties paid.

One Way to Study Char

I thought about Jack Dean many times, and was sad to learn that he passed away only a couple years after I interviewed him. I think of all I learned from him about various species of char. I think especially of his technique of angling as a means to sample a water and learning something about fish. In 2022 I am fishing near the stream mouth of a beautiful alpine lake in Lake Clark National Park with Glen

Alsworth Jr, his son Caleb, my friend Phil Brodersen, and Jeff Duck who is one of Glen's guides. It is my last trip to the area as I finish this book. The book is nearly done, I think. I just need to reorganize some and tighten up the prose. Except I keep learning more.

We had intended to spend a day fishing for Dollies and big rainbow trout on a river down in Katmai National Park. High winds and high water conspire to prevent that. So we made our way back north to Lake Clark National Park. Now I stand on an alluvial gravel fan that drops off very quickly into deep water, seventy-five yards from where float plane is parked and ten yards from the mouth of the stream. I have just caught several small but beautiful Arctic char. Their spots range from pale pink to white. Their underbellies and lips both have just a faint orange tint. None are over a foot long. I am casting flies imitating sculpin, largely because of what I had learned from Jack Dean about char diets many years early.

Glen tells me there is another population of Arctic char in the lake that grow much larger than the ones we are catching. A biologist has told him the larger ones have become piscivores rather than insectivores, which accounts for the difference in size. I'm seeing now that the char in certain waters in the Bristol Bay drainage exhibit the same phenomenon as in the lakes Jack Dean studied on the Kenai Peninsula. One part of my mind files that information away as having potential interest for the book. The other part of my mind—the one more dominant at the moment—has a more immediate and practice use for the information. Or, rather, I decide it's time to try one of Jack Dean's research methods.

"There are bigger char in this lake?" I ask Glen. Glen repeats what he had told me early about two separate populations: the dwarf insectivore char and the larger piscivores. "Where are the bigger ones?" I ask him. He opines that they might be out farther and down deeper at the moment. I walk to the mouth of the stream. Though the creek is small, the current is swift. Swift enough that crossing it had not been easy. I tie on a larger streamer, still dark but a little flashier: a Dolly Llama. I have on some sinking butt leader. I cast my fly out fifty feet into the lake across the current, and let the current take it another fifty feet to the backing, quite a bit past the limit of my casting ability especially with the sinking line and heavy fly. I don't know how deep the water is, or how fast my fly will sink. I guess and starting counting slowly to twenty. Then I begin to strip.

A brown bear strips the eggs out of a female salmon. When sockeye are spawning in the rivers, they are often so abundant that bears will ignore the flesh and strategically focus on the most protein-rich parts of the fish—eating the eggs from the female salmon and the brains of the male salmon.

Page 65: Two members of the genus *Salvelinus* native to Alaska and common in the Bristol Bay drainage. (Top) A male Dolly Varden char (*S.malma*) from the Koktuli river in late September, and (bottom) a lake trout (*S. namaycush*) from Kulik Lake in late August. Many Dolly Varden char have an anadromous life history and can be found following runs of spawning salmon to feast on eggs. Lake trout have lower tolerance for salinity and are generally lacustrine, but there are diadromous populations that take advantage of abundant food in brackish estuarial waters.

Page 66: (Top) A male sockeye salmon (*Onchorhynchus nerka*) in characteristic spawning colors and with its famous pronounced kype. Sockeye salmon are the most abundant salmon species in the Bristol Bay drainage vital both ecologically and economically, and continue to thrive with healthy populations. (Bottom) A male silver salmon (*Onchorhynchus kisutch*), also in fall spawning colors. Though less abundant than sockeye, silver salmon also play an important ecological role, and are a popular target among sport fishers. Silver salmon are known to feed on juvenile sockeye—a habit once wrongly blamed on Dolly Varden char.

Page 68: Fall colors, glacial ice, and a dusting of new snow in the Chigmit Mountains, late September.

Page 69: (Top) A brown bear sow gives a ride to her spring cub amidst a field of fireweed. (Middle) A brown bear prepares to feast on her freshly caught sockeye, while another ponders whether to move in for a steal. Seagulls can only hope for the scraps. (Bottom) Mulchatna caribou herd in early August with their antlers in velvet.

Pages 70 - 71: A red fox and a willow ptarmigan near Moraine Creek, early September.

Page 72: (Top) Lake Kontrashibuna seen from the air, looking southeastward from the outlet toward the Alaska Range, with the southern slope of Tanalian Mountain visible to the left. (Bottom) One of Glen's Cessna's sits in front of the Farm Lodge (Lake Clark Resort) on Hardenburg Bay on Lake Clark in the town of Port Alsworth surrounded Lake Clark National Park.

Part II:
Days on the Water:
Experiences of
Abundance and Diversity

4.
The Koktuli and the Chulitna: Of Footprints, Rivers, and the Start of a Friendship

As I finish putting this book together—many years after starting it—I think back on my first meeting with Glen Alsworth Jr. in the summer of 2015. I had received a small travel grant to support writing and research in Bristol Bay. My good friend Dave O'Hara, with whom I had co-written three previous books, would be joining me. Though not the only reason for our trip, the proposed Pebble Mine was an important focus for both of us. We had read reports about the mine including: the geology and hydrology of the area and the permeability of the soil, the size of the proposed dam for storing toxic mine tailings, and possible locations and impacts of the haul road for getting the ore out. We'd also read about the likely impact on the native cultures of the Dena'Ina people who had dwelt in the region for centuries. It would be an understatement to say that none of what we read made such a massive open-pit mining project in the headwaters of the world's most important and productive salmon water seem like a good idea. But we wanted to visit the area and talk to some people who would be impacted (perhaps positively or negatively). Dave and I were co-teaching a nature and environmental writing class in Alaska that summer for Middlebury College, with a focus on rivers, ecology, and Dolly Varden char. The class was taught in the Anchorage area, up the Matanuska Valley, and on the Kenai Peninsula, but the topic was great preparation for a Bristol Bay trip.

When I reached out to Glen in early 2015 to book a visit, I knew little about The Farm Lodge other than its location and the information I could glean off the lodge website. It was largely an instinctive choice. My grant budget was enough to cover round trip from Anchorage to Port Alsworth, a short stay at the lodge, and a one-day fly-out to the vicinity of the proposed Pebble Mine. The rest of our visit to Lake Clark National Park and Preserve would be do-it-yourself-on-a-budget. At Glen's suggestion we planned a three-day hike-in camping trip to Lake Konstrashibuna, a little over three miles upriver from Port Alsworth, for after our lodge stay.

I'd corresponded with Glen a little bit in the spring when I booked the trip, but we only met in person that August morning standing beside his plane on the dirt airstrip in front of the lodge as we prepared for our one fly-out expedition. His demeanor seemed serious as he went through the standard safety precautions before flying. We hadn't been in the air long, however, before his quiet humor emerged. Although I was mostly looking forward to time with Dave, I could tell I would enjoy hanging out with Glen for a day even if our paths never crossed again.

Because of our interest in the Pebble Mine, Glen had suggested a fly-out day to Koktuli River. If the weather was cooperative, we could even do a flyover of Frying Pan Lake and the mine's base camp on our way back to Port Alsworth. Emerging right from "ground zero" of the mine site, the Koktuli would be the river most immediately (and adversely) impacted by the mine, the dam, and the tremendous disturbance to the area's hydrology. The South Fork of the Koktuli (often referred to simply as the Koktuli) has its upper headwaters in various springs and wetlands on the southern slope of the hills in the mine's proposed footprint, with a major source coming straight out Frying Pan Lake where a mining base camp had already been set up and where the massive tailings dam would be built. After flowing south out of Frying Pan Lake for just a couple miles, the river turns west. Gathering up more waters draining south out of the hills in the mine's footprint, it bends northwest, picks up a couple larger rivers including its north fork, and flows into the Mulchatna River. The North Fork of the Koktuli starts right at the center of the planned footprint where the mine would turn the landscape into a massive open pit. It flows north and then west, draining the opposite side of the same hills that feed the South Fork, before turning south at the western end of those hills, and then west again to join its sibling. The Mulchatna then carries their combined water southwest into the Nushagak River, which continues south into Bristol Bay at Dillingham. Thus, to say that Bristol Bay is downstream of Pebble Mine is true both metaphorically and in a literal geographic sense.

Though it didn't dawn on me at the time of Glen's suggestion, visiting a river on the east side of the Nushagak drainage would also be an appropriate bookend to the trip with my father on the west side of the Nushagak drainage a dozen years earlier. I had little idea at the time that the 2015 trip would be the first of many to Port Alsworth over eight seasons from 2015 through 2022. I also had little idea how many ups and downs the proposed Pebble Mine would take, with the project repeatedly getting killed (seemingly), resurrected, and then wounded and revived yet again. I also had little idea how many other important stories would emerge. Or that the day would be the first of many I would enjoy spending with Glen in years to come, and that through Glen I would get to know both Branden Hummel and Jeff Duck (guides for The Farm Lodge) and Liz Davidson (one of the pilots for Lake Clark Air) as well as several other pilots and employees of the lodge, and others of Glen's family members. And as I got to know the area better, my visits to Port Alsworth would also include camping trips to both Lake Clark National Park and Preserve and Katmai National Park and Preserve. One fall I would come and teach workshops for the public school in Port Alsworth, spending several days in town in a different context. I would visit in October one year and speak at the local church, and then return each of the next two years to speak four more times. Another year, I had a thoroughly enjoyable afternoon flight and fishing excursion with Don Welty who at the time was the pilot for Lake Clark National Park and Preserve (though it was his day off, and our trip was in his personal float plane, a little Piper Super Cub). And I got to know the park historian John Branson whose books and presentations would inform my writing, and whose friendship would make me a better person.

But it was Glen I would get to know the best. I would come to look forward almost as much to our times of conversation in his airplane flying over the landscapes of Bristol Bay as I would the actual fishing. (And I looked forward to the fishing a lot!) I appreciated his thoughtfulness, his sense of humor, his integrity and care for his employees and family, and his deep his knowledge of the land—the knowledge of somebody who had lived his whole life in and over that landscape, whose parents, grandparents, and children also lived (or had lived) there, and who cared about the land.

In addition to several visits to the Koktuli River, over those eight seasons I also traveled to rivers and lakes high up on the drainage of the Chilakodrotna—another headwater of the Nushagak. Glen also took me (or sent me with other pilots and guides) to multiple lakes and rivers in the Lake Clark drainage both north and south of Lake Clark, and to other waters farther down the Lake Iliamna drainage. We visited numerous waters in Katmai National Park and Preserve including on both the Naknek and Alegnek drainages. And each new water we visited, as well as each time I returned to some favorite spot, increased my sense of the beauty and grandeur of the landscape and its waters, and of the important ecology diversity of the Bristol Bay drainage. Each visit also deepened my feelings of *hiraeth* and *sehnsucht*. And while I visited only a tiny fraction of the thousands of rivers, streams, and lakes in the drainage, even that small fraction of the drainage's waters showed a tremendous amount of diversity.

The remainder of this book recounts some of those experiences and descriptions, almost all of which involved the casting of a fly rod. Though in some cases the details of the locations have been omitted, I have sought to provide an accurate description of the rivers, landscapes, and my experiences of them. Indeed, when it comes to accuracy, the experiences are among the rare ones in my fishing life for which I don't need to exaggerate the size of the fish I caught. Or the vastness of the landscape, the intensity of the wind, the beauty of the settings, or the nearness of the bears.

The South Fork of the Koktuli River (and the Pebble Mine)

When the fat tires on his black Stinson roll to a halt, Glen Alsworth Jr. climbs out from the pilot's seat. David O'Hara and I emerge after him into the damp morning air. We stand a moment beside the plane at the end of a long stretch of gravel on a low flat ridge—a natural landing strip in the middle of tundra where Glen has brought us safely down. Low rounded hills rise to our north. To the south, the land falls away toward the vast gray-blue quilt of Lake Iliamna. A handful of evergreens taller than I am poke out of the tundra in a sheltered hollow a few hundred yards away, but for miles around the land is covered mostly with lichen and moss, scattered patches of grass, wildflowers, low-lying shrubs, and dwarf trees that barely reach my ankles. And, as we will soon learn, mingled with the moss and grass are myriad wild blueberries and crowberries, as well as a few lowbush cranberries. It's mid-August and the blueberries and crowberries are ripe. It's a beautiful scene, though in other ways a stark and lonely one as well, especially with the damp breeze and overcast sky. I'm also a bit confused about exactly where we are and why we've landed. Although this is my first trip to the Koktuli, I'd been studying maps of the region for weeks getting ready for the trip. For the last several minutes, the landscape we were flying over had not matched up with my mental map of where we were headed, and—unless I'm completely mistaken—the plane had circled a couple times before landing. There is no river within sight.

"The cloud ceiling is too low for us to get through the pass right now," Glen explains. It is my first experience of just how seriously Glen and his pilots take safety concerns. "I don't want to keep burning fuel, so we'll just wait on the ground for the ceiling to rise another hundred feet. Probably an hour or so."

I nod. I try not to be impatient, but I'm eager to be fishing. Although I'd taken the float trip with my father a dozen years earlier, this is the first fly-out fishing trip of my life. It's also the only one planned for my writing and research trip to the Bristol Bay drainage. (At the time, I'm thinking about writing just a single essay or article, and not an entire book.) Each minute is precious. I can feel the seconds ticking. Yet there isn't a thing we can do to speed up the clouds. Glen reaches into the plane, pulls out a large plastic Ziploc bag, and strolls off toward the soft tundra at the edge of the gravel where he proceeds to pick wild berries.

I take a few photos of the landscape, berries, lichen, moss, dew-covered fireweed, and some pretty mushrooms. Then I help with the berry gathering. By the time the clouds have lifted enough to fly through the pass, Glen's bag is bulging with a mix of berries.

We're soon in the air again, and this time we're able to get through the pass safely. Before long, we're flying low over the South Fork of the Koktuli. Below, I see a brown bear splashing hurriedly across the river. I assume it's chasing salmon. I snap a photo out the window. Only after I get home and download the photo to my computer where I can look at it on a larger screen will I see the family of three black bears in the brush above the river, moving away from the pursuing brown bear. When the Stinson drops back down on the tundra, we are on a bluff beside the river. No plastic bag and berry-picking this time. The three of us are out in a moment rigging up our fly rods. We are after Dolly Varden and maybe rainbows. There are also sockeye salmon in the river, so I tie on an egg-sucking leech.

After briefly spotting what he thinks is a large Dolly from atop the bluff, Glen directs Dave and me to two good locations on the same long hole. Dave climbs down a steep bank to the middle of the pool, and fishes the lower end where the Dolly might have swum off to. After taking several photos of Dave, I move to the top of the run to drift my fly into the deep water on the far bank where Glen first spotted the Dolly even though he can no longer see it there. I cast where he tells me. And I quickly learn that 3x tippet, which seemed heavy to me after years fishing little New England trout streams, is not heavy at all for Alaskan trout water. The big Dolly that has sucked in my egg-sucking leech looks at my 3x and laughs. I hear the laughter reverberating off the bluff as it shakes it shoulders and swims off with my leech.

Fortunately, it will not be the only Dolly who takes a leech that day. The first one I land, fifty yards upriver from where we started fishing, is a little over twenty inches long. Unlike the smaller southern Dollies I'd caught in mountains of Chugach State Park or up on the Kenai river miles above tidewater, whose sides were light silvery green, and whose spots were a pale magenta, this Koktuli River fish has vibrant green sides and spots that are almost blood-red. *Cera* red. The intense green makes those red spots stand out even more. But she has no red on the underside. A white belly and a torpedo-shaped head lacking a kype tell me it's a female. I'm enamored by the beauty of the fish as I release it back to the wild. Although it's likely that three fish I caught earlier in the Iliamna River that had been identified by my guide as Arctic char were actually northern Dolly Varden, this is the first fish I catch in Bristol Bay that I *know* to be a Dolly as I am landing it, holding it, and releasing it.

We work our way a few hundred yards upriver from where we land, and then a few hundred yards downriver. After Dave and I have both caught some fish, Glen begins to fish with us. We catch several Dollies between us, and even more grayling. We see numerous sockeye salmon, mostly in the deeper channels by the shore as they slowly migrate upriver in schools. We also spot two or three Chinook salmon, distinguishable from the sockeye by their larger size, more streamlined bodies, and their coloring. (While both Chinook salmon and sockeye salmon turn red late in their spawning runs, the Chinooks are typically red all the way up through the head whereas the sockeye get dark green heads.) It is a delightful day in every respect.

On the flight back to Port Alsworth at the end of the day, we fly over the base camp for the proposed Pebble Mine. Although the mine has not yet been approved, preparation work is clearly going on. Shipping containers, heavy-duty tents, small portable houses, and what appear to be out-houses, steel girders, corrugated metal, and pallets of other supplies are stacked or spread over an area of a couple acres. I also notice what I think is a helicopter pad. And arcing out like tentacles from the supply depot are the scars and tracks where ATVs or other equipment has chewed up the tundra. The proposed mining plan at the time includes one of the longest and tallest earthen works dams on the continent that will hold a lake of toxic heavy metal mine tailings. I try to imagine what the landscape will look like if the dam is approved. I picture the path water will take down if the dam leaks, spills, or breaks: down the Koktuli where I was just fishing, on into the Mulchatna, then down the Nushagak and into the salmon-rich waters of the Bristol Bay. I also think back on the reports and studies of the hydrology of the area: the porous gravel that lies beneath the tundra, and how the toxins will seep through that even if by great fortune there is never any breach of the dam itself. I think of how dramatically such a large, deep open pit mine will impact the hydrology of this watershed. I think about what the river we just spent the day on will be like if the mine is approved. After several minutes of this, I find myself trying to unthink those thoughts.

South Fork of the Koktuli, Revisited (Again and Again)

Two years later, in 2017, Dave and I return to the Koktuli on almost the same date. We travel in two planes this time. Dave's son Michael is with us along with guide Branden Hummel. The plan is for the pilots—one of whom is Glen—to drop the four of us off for the day. Once again, we are delayed as we land on the tundra somewhere south of the Koktuli and pick berries while waiting for the cloud ceiling to rise. Instead of seeing brown bears chasing black bears across the river, we spot a heard of eight caribou grazing on lichen.

And then, whether it's better conditions or more experience, Dave and I manage to land even bigger and more beautiful Dolly Varden. The largest of the day are male. Their varying shades of green and red are beautiful and flamboyant to the point of transcendence. Emeralds could not be more green than the sides of these fish. Their spots, rather than the palish magenta we are used to, are fluorescent rubies. Their underbellies range from sunset orange to maple-leaf red. Or red like the blood that drips from a spear wound. And the lips. The lips! They are something to behold.

The day is splendid. Although I've had numerous fishing days with more fish, I've never had a more enjoyable day casting flies, nor have I caught any fish more beautiful. I also pay less attention to the Pebble Mine area as we fly back over it. The EPA had denied rights to the mine. It appears that the proposed project is now officially dead. One of the leading conservation organizations is convinced of that, anyway. It has turned its attention from opposing the Pebble Mine to another potential project that also has potentially disastrous environmental consequences impacting salmon: a large-scale hydro project on the Susitna River north of Anchorage. I am skeptical the Pebble Mine project is *really* dead, but I'm hopeful. Anyway, some of my own attention has turned to the impacts of climate change on Alaska: the melting of permafrost, increases in intensity and duration of the wildfire season, flooding of coastal villages, and disruption of wildlife and fish spawning patterns. I don't know how quickly the Pebble Mine situation is about to turn around again.

Yet another two years pass before I next return to the Koktuli, this time with my friend Rich. It's the exact same date as the previous trip, right in the middle of August. The conditions are noticeably different, however. It's been a hot, dry summer. The sky is crisp and blue. We don't have to wait for the cloud ceiling to rise to get over the pass, but we do fly over some smoldering fires. I've been reading about fires all over Alaska. All three locations where Rich and I planned visits—Denali State Park on the Susitna River drainage, the

Kenai River, and some of the Bristol Bay watershed in and around Lake Clark—have been impacted by heat and fire. Our trip to the Kenai River is canceled altogether when the only highway there gets closed and evacuation warnings are sent out.

Our trip to the Koktuli is not canceled. However, the water level is very low. It's low in rivers all over the Bristol Bay region. So low that salmon have not yet come up some of the rivers where they are usually found spawning by this time. Stretches of the Koktuli I've fished on previous trips are too shallow to fish today. The Dollies we had come for are nowhere to be found.

Glen is our pilot again. This time, he plans to spend the day with us. The three of us start out together, but when a warm wind kicks up Glen has to go back to keep an eye on the plane. Rich and I spend the day catching rainbow trout and grayling. Though it's my third trip to the river, they are the first rainbow trout I've caught here. We manage to land quite a few of them between us, all on my go-to egg-sucking leech patterns. With the river running so low, the bows are congregating in just a couple deep holes below a gravel riffle where the sockeye salmon are stacked up to spawn. But once we find them, we see plenty of action swinging leeches over the lip of riffle or beneath an undercut bank on the outside bend. We also manage to land a couple sockeye salmon on our 5-wt rods.

It is only at the end of the day that I manage to pull out a single Dolly Varden. The fishing has slowed down in the afternoon. Maybe we've caught all the bows out of the few places we've found them. Or maybe the bright afternoon sunlight has them even more spooked. Or maybe they've just seen all my egg-sucking leech patterns. So I swap out my ESL and experiment with a new fly of my own creation that I dubbed an *egg-sucking worm*. Though inspired by the egg-sucking leach, it is much smaller. I tied it on a heavily weighted jig hook rather than a streamer hook, and with a small rubber tail rather than marabou. It has no chenille body. No body at all, in fact. Just the rubber tail and an egg head tied with several wraps of sparkly pink chenille. Though I'm fishing the same hole where we'd landed several rainbows earlier, this fly is too heavy to swing under the undercut banks. So I just let it drop over a lip and down into a deep dark well toward the tail of the pool where I'd not caught anything earlier. That's where it stops moving. Hoping it's a subtle take and not a snag, I lift my rod swiftly. The Dolly I land a few minutes later measures twenty-seven inches long making it the largest I have caught outside the Arctic.

In September of 2020, I'm back on the South Fork of the Koktuli for a fourth time. It's my first time getting to fish the river in back-to-back years. The off-again on-again Pebble Mine is now on again, having been at least temporarily resurrected with an overturning of the previous EPA decision. But so has the opposition to the mine, including from some unexpected places. Not wanting to leave such an important decision to the whims of any particular administration, however, conservation organizations are taking some new approaches. A collaboration is in the works between four organizations: the Bristol Bay Native Corporation and Pedro Bay Corporation (two corporations representing native groups), along with Bristol Bay Heritage Land Trust and The Conservation Fund. They will work toward purchasing conservation easements that would restrict development and prevent haul roads from being built into Pebble Mine. It's a $20 million project. As I write this sentence, fundraising has reached 75% of that goal.

On this trip to the Koktuli, it's just me and Branden Hummel. The pandemic has resulted in the Farm Lodge closing early. The last of their regular clients left the day I arrived. Mid-morning Glen drops Branden and me off by the river, and then takes off to do some other pickups and deliveries. It's moose-hunting season so despite the closure of the lodge the pilots have had a few busy days flying all over the Bristol Bay drainage. I'm not sure what to expect this time. Not only did the river fish very differently in 2019 than it did on the same date in 2017, but now I'm here a month later in the season. We're past the sockeye salmon spawning run, but silver salmon have moved up the river. Also, shortly after the plane lands I discover another thing that will impact the day: my fly rod is still on the deck of my cabin. I'll get only half as many casts as usual—and I'll only get that many because Branden generously offers to let me take turns with his rod. The good news is that I'll get a lot more photos than usual. Whichever of us isn't using the rod at a given time becomes the fish spotter and gets the camera.

And the day turns out to be a lot of fun despite the rod shortage. With just the two of us, forced to stay close, I get to know Branden a lot better. We work our way up through the usual spots. Initially we see nothing. Not even the deep honey hole yields any fish. Several bends upstream, however, we come upon the silvers. They are paired up in the shallows in places I hadn't found Dollies or rainbows before. Not long after spotting the silvers, we begin seeing the Dollies. As expected, they are sitting behind the silvers in calf-deep to knee-deep water. Some are out in the open, but many are tight to an undercut grassy bank. They aren't as large as the Dollies I have caught in the past, but there are many more of them. If we're careful in our movements not to spook them, we can sight-fish. We do this with only modest success for quite some time with a large arsenal of flies. We throw at them every size, color, and variety of egg-sucking leech and egg-sucking worm that we have, along with some other patterns. The fish are there, and obviously feeding—almost certainly on eggs—but they just aren't taking our offerings. Branden lands a single Dolly on an egg-sucking leech. Mostly they respond to our flies by moving out of the way when they drift past their noses.

Only late in the morning when we start drifting small, individual egg imitations past them do they start taking. And then it is fast action for the rest of the day. None of the Dollies are over two feet long and many are closer to sixteen inches—smaller on average than we've seen in the past.

Although a couple have some girth, one or two are downright snaky. This surprises me. In the Bristol Bay drainage in September they should have had a couple months of gorging on eggs. The males still sport their bright lipstick and bellies, but their backs and sides are darker and more olive in tone, and not bright green as I've seen here in the past. The smaller run of salmon thanks to lower water levels has had an impact on these fish.

When we've caught enough Dollies, we start fishing for the silvers with Branden's five-weight. I land a gorgeously colored male, slightly bigger than the Dollies we've been catching. It has a big scar on one side, evidence of a narrow escape from some predator. Bear? Eagle? We can only guess. We are reminded of how many creatures feed on the salmon. A hundred yards downriver, Branden lands a silver at least double the weight of my scarred fish—one that would have put a big bend even on an eight-weight rod.

And then we come to September of 2022. Glen says neither the rainbow trout nor the Dollies have been in the Koktuli at all. Heavy rains have caused a lot of flooding, which has washed all the food away: eggs as well as decaying salmon flesh. Without the abundant protein to gorge on, the trout and char have dropped out of the upper river. *To where?* I wonder. Unlike the rivers we have fished in Katmai or on the west side of the Nushagak, the Koktuli system has no big lakes for them to drop down to. To find big water, they would have to swim all the way down to the Mulchatna or even into the Nushagak. I couldn't imagine a

New England trout moving that far to find food or water. But then I think of what Daniel Schindler had recently told me about the long distances that predators—including species of trout and char—go following the spawning salmon. The Dollies could be anywhere in the system. Maybe even back in saltwater. Maybe they'll come back up if the silvers come in. Glen thinks it's possible. But I have only three days to visit, and it's not worth burning a day to find out.

Chulitna Bay Northern Pike

It's 2017, my second trip to Port Alsworth. I'm back with my friend Dave. His son Michael is with us. It's a more extensive trip this year, and I'm excited. In addition to a return outing to the Koktuli, Glen is going to take us down to Katmai National Preserve to a tundra river on the Naknek Drainage. Before either of these trips, though, we get a boat trip across Lake Clark to cast for northern pike (*Esox lucius*) with Jeff Duck near the mouth of the Chulitna River.

I didn't come to Alaska to go pike fishing. I say that as a guilty confession of my inadequate knowledge. It isn't that I don't enjoy pike fishing. It's just that I had char and trout on the mind, and I haven't associated the area with pike. Perhaps because I haven't yet seen any of the pike-fishing videos that portray Alaska as being as much a paradise for pike as it is a Shangri-La for salmonids. I will see those videos later, after this trip. Not that I will need to. Thanks to my 2017 trip with Jeff Duck, my perspective will quickly change even without those videos; chasing pike with a fly rod will become a significant objective in later trips to Bristol Bay. This year, however, the pike feel peripheral.

What is less peripheral is that many headwater tributaries of the Chulitna River emerge from the area of the proposed Pebble Mine. Though the main branch of the Chulitna begins farther north at Nikabuna Lake, some of its waters arise within meters of headwaters of the North Fork of the Koktuli close to the footprint of mine. Although both forks of the Koktuli eventually flow west and reach Bristol Bay via the Nushagak River, myriad little streams also flow north out of the mine area into the Chulitna River, and from there the water flows east and back south down into Lake Clark, eventually reaching Bristol Bay in a much more circuitous route through the Newhalen River, Lake Iliamna, and the Kvichak River.

Chulitna Bay, where the Chulitna River enters Lake Clark, is long and shallow with scattered weed beds ready to snarl propellers of unwary boaters. At the far western end, grass and weeds stretch across the width of the bay, and large flocks of waterfowl cruise the area and nest around the small islands. Though the area is vibrant with life and full of its own type of beauty, the mountains that line the lakeshore to the east and south fall away when they reach the bay making the scene seem less dramatic. Along its final three miles, the Chulitna river is braided in multiple channels as it meanders between grassy banks through a wide flat terrain with little noticeable current. Unless they know where to look, visitors might not even notice the river mouth until they are close.

After a half-hour ride from the lodge, Jeff Duck cuts the engine and steers the boat onto a beach. There is nothing marking the spot as good fishing water, except perhaps the weed bed reachable from shore by a modest cast. One thing I do notice, though, is that the water is much clearer here—less green than the main part of the lake, with much less glacial flour. I will later attribute this to the influx of water from the Chulitna River, which is not glacial.

Since I didn't come expecting to fish for pike, I have not brought any wire pike leader. The best I can do is use 1x tippet. I also don't have any pike flies with me. These are both mistakes that I will rectify by my next visit after learning how good the pike fishing is! Meanwhile, however, Jeff suggests we try our largest streamers or leechiest looking flies. He has a couple we can use, but I stick with my own arsenal even though the hook sizes are too small for pike. Jeff directs us to cast across a channel to the edge of the weeds, and also not to neglect the near shore or the bottom of the channel just beyond our rod tips.

My first catch of the day is a two-foot long pike. It takes a big black leech that I let sink down to the deepest part of the channel. The take and the fight are fun, but I am not overly impressed; I can catch two-foot-long pike in several ponds and rivers near my house in Vermont. Admittedly the scenery is a more spectacular here, especially if I look off in the distance toward the higher peaks. I also appreciate that I'm not fishing with several other boats around me, or the noise of passing trucks, as would likely be the case casting for pike in Vermont. Still, I've caught trout bigger than this pike. Little do I know that this first Lake Clark pike is the smallest one I will catch here over several years and trips.

My next pike, half an hour later, is three feet long. I try to act nonchalant, but this second one *does* feel like a big fish. It's bigger even than many west coast steelhead I have landed. My little trout fly is just a tiny speck somewhere deep in an intimidating mouth full of long sharp teeth. I'm glad Jeff comes over with his long forceps to help extract my hook and save my fingers. Although I have pike-fishing friends in Vermont who break the forty-inch mark a few times each summer, this is bigger than most anything I see. It's also a good trend: a two-foot fish, then a three-foot fish. I hope the pattern keeps up.

My third pike is not a four-footer, but it's close. It takes my fly like the ambush predator it is, torpedoing across the surface with a visible wake that has my adrenaline pumping even before it hits my fly. The fish, which takes some time to land, is forty-five inches long: far and away the biggest pike of my life, and as big as the largest caught in a typical year by my top-notch pike-fishing friends—the ones who were disgusted to learn I had gone to Alaska without pike gear.

I will return to this spot many times in the next few years. It is only two trips later that I hit the four-foot mark on an epic September evening with Branden Hummel. In two hours of golden evening light I land eight fish, two of which are over forty inches, and one of which absolutely torpedoes a surface fly. Although we release all the fish, we run the length and girth of the big one through a couple formulas which estimate it to be between twenty-seven and thirty pounds.

Later, I will ponder the relationship of pike with the ecology of Bristol Bay. There are places in Alaska including on the Kenai Peninsula where introduced non-native pike have had devastating impacts on native fish including Arctic char and salmon. In those places, it is illegal to return a pike live to the water; all caught pike must be kept or killed. In Lake Clark, however, they are native, with regulations in place to protect them from fishing pressure and over-harvesting. And the pike fishery really can be described as world class. (Two years after I caught the four-foot pike with him, Branden guided a Farm Lodge client to a pike over fifty-one inches.) Most of Lake Clark is cloudy with glacial flour, and the further east one goes toward the higher peaks, the cloudier the water is with the color turning from light turquoise to bright green to sludgy gray. Closer to the inlet of the Tlikakila River, visibility in the summer isn't much more than a foot and a half. With so little light reaching the bottom of the lake even in the shallow areas—so little of the sun's energy to be turned into chemical energy through photosynthesis—there is little plant life to form the bottom of the food chain. As a result, Lake Clark is not nutrient rich. Reports state that the lake contains all three species of char native to the Bristol Bay drainage (lake trout, Arctic char, and Dolly Varden), as well as rainbow trout, grayling, bourbut, and pike. But it doesn't contain *any* of these species in great numbers. I have caught numerous grayling at the mouth of the Tanalian where it flows into Lake Clark, but over dozens of hours fishing in waters reachable by fish from Lake Clark I have caught only a single rainbow trout and a single lake trout.

Yet in certain parts of the lake, pike thrive. They are abundant, and they grow large. These top-of-the-food-chain predators thrive because food is abundant. For the most part, these "certain parts of the lake" are shallow coves where clearer rivers flow in and where the sun reaches the lake bottom. They are also farther from the glacial inlets, toward the western end of the lake where more of the silt has had a chance to settle. Between the shallower depth and the clearer water, these bays allow abundant plant life supporting a whole complex rich food web with numerous waterfowl and plenty of forage fish. There is no shortage of nutrients.

Of course, there is also another source of protein that does not come from the plant life at the bottom of the food chain. Most summers, approximately 500,000 sockeye salmon come from Bristol Bay up into Lake Clark, bringing with them a few million pounds of marine-derived nutrients. Much of this MDN remains in the lake or on the surrounding landscape. Many of those salmon make their way into Chulitna Bay and up the Chulitna River. I am told that pike feed not only on salmon fry, but even on adult salmon. Although I wouldn't normally think of a large ocean-going salmon entering a freshwater lake to become prey of another fish, I have at times been fighting a Lake Clark pike well over 30 inches long and watched an even bigger pike swim in and attack the fish I was fighting. I have no doubt that a four-foot-long pike would gladly follow a school of salmon around just like everything else does, and attack even a healthy adult salmon. At least one person I know who does subsistence salmon fishing in the lake with nets has caught an occasional monster pike swimming in among the salmon. Those MDN impact the entire ecosystem.

5.
Katmai National Park and Preserve: More Tiles in the Mosaic

Even if I had never set foot on the ground of Katmai National Park and Preserve and cast flies into any of its rivers or lakes, the flights from Port Alsworth across Lake Iliamna and down into the protected parklands—flights I have had the privilege of enjoying on several occasions in one of the Farm Lodge float-outfitted Beavers or Cessnas or fat-tired Stinsons—give an appreciation and awe for the diversity of the waters in this drainage. Looking down from the plane at bears, caribou, beavers, swans, moose, and schools of spawning sockeye salmon so thick and large they can turn entire long pools of the rivers red, also offers a glimpse of abundance and of the importance of that diversity. So does flying over the native villages that dot the landscape, reminding me of peoples who have known and depended upon this land for generations.

And each flight gives me as well a longing to drop down on some of those rivers and explore their diversity of fishes, waters, and landscapes with the aid of my fly rod and a few flies. I am profoundly thankful for the few of those rivers I have visited, and for the colorful tiles I have seen up close that help present that mosaic image of Bristol Bay. Even knowing that I will never see the majority of the rivers, streams, and lakes (except from a few hundred feet up in the air) —and that even among those rivers I *have* had the opportunity to visit, I have seen only a small portion of them—I am content to know they exist.

The Alagnak Drainage: of Big Winds and Big 'Bows (and also more Dollies)

It is August of 2017. I'm on a fly-out trip with Glen mostly for the pursuit of rainbow trout and Dolly Varden char, but the brown bears will make an equally lasting impression. As we fly low over Katmai National Preserve along the Alagnak drainage, we see several bears fishing in the rivers below us or wandering along the banks. We also spot a small herd of caribou splashing across a shallow river. Even from the air, the bears look big. From the ground they appear even larger.

The general area where Glen brings Dave, Michael, and me is a popular destination for both anglers and bear viewers. A lake, large enough to land a float plane on under most conditions, makes a convenient drop-off point for one-day fly-out wade-fishing excursions as well as a launching point for float trips. Planes also bring bear viewers from as far away as the Kenai Peninsula. I can see why. Two years later I will return with my wife specifically to watch bears (although I will get some fishing in, too.) At one point five of them will be visible spread out upriver, downriver, across the gravel bear, and—most excitingly—just a few dozen yards away directly below the low bluff where we sit. A year after that, I will bring my son Mark and daughter-in-law Ellie and we will see twenty-one bears in a single day including three sets of sows with triplet spring cubs. A three-year-old will walk right past us down the river, just a dozen yards away. It will give us just a single intimidating sideways glance to make sure we are staying in our place up the gravel bar and yielding the river (and its salmon) to him.

My level of concern will ratchet up a step when a sow with triplets appears on the bank above a pool just twenty yards downriver. "Shouldn't we be concerned about a sow with cubs?" I'll ask Glen. "I mean, aren't sows notoriously defensive about their young?"

Glen is not flippant about bears. He takes them very seriously. Long before any of our encounters, he has gone through safety protocols as well as strategies to minimize our impact on the landscape and its ecosystem. These include not setting anything down that might get forgotten (or left behind in a hurried departure), not doing anything that would teach bears to associate humans with food, and also staying together so that if a bear ends up wanting to move in our direction in a hurry (for example, to move away from another larger bear) it only has one clump of humans to avoid. But though Glen is careful and cautious, he is not fearful. He explains that sows with cubs like to be near people because it's safer for their cubs. A momma bear's biggest concern is not a 200-pound adult human, but an 800-pound male brown bear. Females can nurse their young for up to three years. Male brown bears will eat even their own offspring in order to get females to stop nursing so they become estrous again. However, the big males tend to be warier around humans and to avoid them altogether. So sows will keep their cubs close to anglers and bear viewers—a benefit to the viewers, though not necessarily to the anglers.

It's the angling that has prompted my trip. Fly fishing will bring me to this area seven times from 2015 to 2022 ranging from early August trips when sockeye salmon are just entering the river, to later in August when they are thick and the egg-drop is peaking, to late September when all that remains of the sockeye are their rotting corpses, but a few silver salmon have moved into the river. I will spend far more time in this area than on any other river in Bristol Bay. It will both delight and frustrate me, thrill me with the sensation of watching line zip off my reel as a fish explodes down river, and break my heart (and my tippet) with missed chances. My trips will even include a three-night camping trip. And though three of those seven trips will be with family and students more interested in the bears than the char or trout, the fly fishing is never far from my mind.

Even so, my first experience sitting by a river watching a pair of large brown bears chasing salmon through the shallows only thirty or forty yards away is breathtaking. It takes only a few minutes to figure out which bears are dominant, and to make some sense of the drama between them. The smaller one is the better at catching fish in the shallows, but the bigger one is the playground bully, all too willing to steal lunch from the scrawny kid. This dynamic gets played out in front of us. I can see why the bears are here. Just about every deep pool I look down into is thick with schools of bright red sockeye salmon. We could see them from the airplane like stain in the water. We can spot them from the shore a couple hundred yards away. In the wide shallow stretches where the bears are hunting, we watch one sockeye salmon after another run the gauntlet, darting from the relative safety of one pool to the next, their backs half out of the water, exposed. When a bear catches one, it rips it open looking for eggs. Sometimes they just discard the fish after a nibble or two, like a kid who is served some undesirable vegetable. And it's not just bears feeding. Seagulls are numerous. Bald eagles line the bluffs. Foxes wander past. On another visit, my sons Mark and Peter and daughters-in-law Ellie and McKenna will see wolves loping along the bluff.

We watch the bears as we eat a leisurely picnic lunch, and for some time after that. Soon, however—as captivating as the bears are to watch—I'm eager to fish again. The fishing has been challenging so far. With so many bears, the fish are easily spooked. Any shadow or movement along the shore sends schools of salmon darting off. And when the salmon spook, the rainbow trout and Dollies also spook. The water we've worked through so far has been shallow and clear, which (like my experience on the Koktuli) makes the fish even more wary. We do spot some big trout. Or, to be more honest, it's mostly Glen who spots them and points them out to us. Though as the day progresses and I get more used to the water and where to look, I've gotten better at it. Some of the trout been holding in knee-deep water under the banks, or in pools behind boulders, but many have been sitting in just sixteen to twenty inches of water in little depressions in the gravel downstream of where sockeye salmon are paired up on redds. It's the 'bows and Dollies in those wide shallow stretches below the redds that seem to be feeding most actively, but are also the ones most easily spooked. In the deeper water, I've tried egg-sucking leeches. In the shallower water I've drifted egg patterns along the bottom. I've elicited a few strikes and a few more looks or misses, but I've spooked more fish than I've enticed to check out my flies. To avoid spooking them, I've started casting from a considerable distance. Fortunately, the water is shallow and clear enough that indicators are not needed, because the indicators also spook the fish. I've so far landed only a single Dolly Varden and one rainbow trout, both long, strong, and fat on eggs. I've also had a couple sockeye salmon angrily snatch my flies, which has cost me several minutes of precious time either fighting the sockeye or rerigging after getting broken off. So when the bears finally clear away after lunch, and I've been able to get some underwater GoPro video of a big school of sockeye salmon, I'm ready to cast more.

In terms of numbers of fish landed, my first day on this drainage turns out to be one of the slower days I'll have with Glen in Bristol Bay. In terms of enjoyment, it lacks nothing. The beauty of the scene and the delight of watching the bears leaves my heart full. As already noted, I will return to this general area several more times over the next half decade. My best day of *catching* fish will come on a later visit when the water is a higher and I can sight fish from far up the bank without spooking them. I will swing dark egg-sucking leeches across the current and watch big shadows torpedo those flies like pike in Chulitna Bay, and I will also entice a couple big rainbows out of deep pools with a bright pink flesh pattern. My biggest frustration with the location is always having to leave in the afternoon.

And this leads to my most recent trip: a three-night camping excursion in August of 2021 with my friends Rich and Rick. Or at least they acknowledged me as a friend *before* I roped them into my hairbrained scheme of camping out on the windy tundra in the middle of bear country on a trout and salmon stream so that we could have all day to fish without needing to fly back to a lodge. In my defense, I did arrange for an electric bear fence, and I didn't know how bad the wind would get on two of our four days there. I also would not have recommended Rich bring a tent that he'd had since his wedding anniversary. Plus, on our Farm Lodge stay prior to our camping trip we'd had some very enjoyable and successful days catching pike in Lake Clark, grayling in the Tanalian River, and lake trout and Arctic char in Lake Kontrashibuna.

After getting our tents set up inside the electric bear fence, we head up to where we've stashed five bear barrels of food sixty yards up the slope away from the tent. We've barely finished eating when a brown bear comes wandering across the tundra in our direction. We quickly close the barrels and slip back inside the electric fence. Not more than three minutes later, the bear wanders over the slope and checks out our food supply. He knocks one barrel over, but when he can't get access to the contents, he loses interest and walks away toward a nearby stream where salmon are spawning.

About then, the wind starts to pick up. It's stiff though not brutal that first day, and the fishing is pretty good. In some rivers feeding high-elevation reservoirs near his home in Colorado Springs, Rick often fishes for Kokanee: a non-anadromous non-native strain of sockeye salmon that have been stocked in various inland waters. We begin referring to the sockeye salmon as "sea-run Kokanee": a very large anadromous Kokanee that will make Colorado anglers jealous. The sockeye are thick in the river. Some are spawning in the main stem. Some are spawning along a nearby lake shore. Some have moved across the lake to spawn in tiny tributary streams.

We fish the afternoon in growing wind. Casting is hard, but we land several fair-hooked "sea-run Kokanee". Rick is having lots of fun with them. Rich and I both comment that the first two or three fair-hooked salmon are quite enjoyable on a 5-wt rod. After that, however, they start to get tiresome for us; in order to cast for rainbow trout and Dolly Varden, we need to fish close to where sockeye are spawning, and we can't keep them off the hook. Even when they aren't attacking our flies, they prove adept at getting snarled in our lines, making it hard to cast for the rainbows and Dollies. Adding to the challenge, the wind rippling the water makes it hard to spot trout. I hook the only two rainbow trout of the day and I land one: an egg-fattened twenty-two inch fish that fights far harder and with more energy than the heavier sockeye salmon.

The second day is similar. The biggest challenge is that several anglers come in by plane and the river is the most crowded of any day I have ever seen it. The best spots on the river get taken by either bears or other anglers, and we spend much of the day playing hopscotch over both types of competition. We land a lot of sockeye salmon, and by late morning we are struggling to keep them *off* our lines. We get broken off by a few trout, but only one rainbow makes it to the net. Today it's Rick's net.

To our surprise, however, we start catching lake trout close to our camp. Several of them take flies near the mouths of two little creeks where sockeye salmon are spawning. The lake trout are about the same size as the rainbow trout and are eating the same thing: salmon eggs. They fight just as hard, too, and (being members of the *Salvelinus* genus) they are even more beautiful and colorful than the 'bows. They help redeem the otherwise tough day.

And then the gale force winds set in along with pelting rain. Casting a fly rod shifts from difficult to impossible. Just standing up straight is a challenge. The wind bends an upraised fly rod almost as much as a strong fish does. Our efforts to cast for rainbow trout and Dollies prove futile. We spend several hours in or near the tent. We watch a sow bear come by with cubs and catch a salmon in the little creek just a few dozen yards from the tent where the lake trout had been feeding, but we are not tempted to join the bear in fishing again.

The next morning, I awake at 5:10am to empty my bladder. It's still dark. I went to bed thinking of a 5:45am wake up to fish a few good spots before any anglers arrive by plane. When I hear rain a few minutes later, however, I turn off my alarm. I don't awake again until 6:45am. Having slept in later than planned proves irrelevant, though. We will see no other anglers or bear viewers that day. And not many trout, either. The east wind is still blowing hard. Too hard for any airplanes to land. So we have the river to ourselves. But it doesn't do a lot of good because it's blowing too hard to cast a fly. The only way to get a fly out on the water more than fifteen feet away is to stand upwind of the target and let the gales strip the fly out. Which is precisely what I do, though it limits the number of places I can fish to just a couple bends. I lose a big bow early on to a poorly tied knot. I blame my cold fingers, but I'm frustrated at myself—especially later when it proves to be the only trout we see all day. By 11am the wind is howling. It's tough just to hold a fly rod up, and fishing in any fashion is impossible. We are back in the tents by 1pm with fierce winds and heavy rains that nearly flatten the tent above us. I keep expecting our tent to sail away. At 6:30pm, I'm still in the tent, leaving the shelter only long enough re-secure stakes. It is a discouraging and disappointing day, leading to a personal record of eighteen straight hours in a tent.

The wind does not die down until the morning of our departure day. We pack up most of our gear, but leave a tent standing. Then we walk downriver. Finally, without the wind and with some brighter light, we begin to see trout.

Maybe we are seeing them better, or maybe they have simply moved up into the river in greater numbers. I spot at least a half dozen rainbows, mostly in little depressions in the shallows. They are feeding, but not taking our flies. We try a variety of egg patterns and egg-sucking leeches with no success. All we succeed in doing is spooking the fish.

We move downriver and try to spot trout from a bluff. This plan comes to an end when we spot a sow with cubs napping just up the slope from the bluff. The trio shows no signs of moving. We think it is better not to end up down in the canyon if she decides to drop down for more feeding. We definitely don't want to get between the mother and the cubs. Anyway, we are running out of time before our pick-up. Feeling defeated, we start back up toward the campsite to await our plane, hoping to have an hour and a half to fish on our walk back for the trout we had spooked earlier.

On the walk back, I begin to notice for the first time of the trip a lot of dead salmon in the river. It's more of those millions of pounds of nutrients brought up each year from the ocean, but in the form of flesh, not eggs. I tie on a smallish pink marabou and chenille flesh fly. Drifting it through a stretch of shallow sandy riffle—water that doesn't look more than ankle deep and has no obvious holding structure—the fly stops suddenly. I set the hook expecting another sockeye salmon. My reel buzzes and my fly takes off downriver like I've hooked a torpedo. Though I have not seen the fish, I know instantly that I've hooked a big rainbow. Several minutes later I land a twenty-five inch fish rippling with fat from several days of non-stop gorging. It's the last fish of the trip and the largest trout we've landed.

The Naknek Drainage: Famous Rivers and Small Unnamed Streams

One of the most famous tributaries of Bristol Bay is Naknek River, which flows through the town of King Salmon. Like the Alagnak drainage—which in places is separated from it by only a narrow ridgeline—the headwaters of the Naknek begin high in Katmai National Park and Preserve. Perhaps the most famous location on the drainage is Brooks Camp. Brown bears are notorious for lining up along Brooks Falls, as many as twenty at a time with some standing atop the falls and snatching salmon out of the air. This is where the annual "fat bear" competition is held every October. My wife spends many summer hours with her computer tuned to the Brooks Camp livestream, watching bears catch salmon. By the time of Fat Bear week, she already knows many bears by name or number.

There are numerous large lakes and rivers in the drainage. American Creek is one of the popular fisheries, and at 1600' in elevation is the highest river I've fished in Katmai. In 2020 I spent a late September day there with Glen, his son Caleb, and Jeff Duck. It was long after the egg drop had ended when the bears and rainbow trout were left eating dead salmon flesh drifting downstream. Flowing out of Hammersly Lake, American Creek follows a big loop northwest and then back south through Coville and Grosvenor lakes before its water enters Naknek Lake and joins the water pouring over Brooks Falls from the south. Naknek is one of those drainages that has tremendous diversity just within the one river system. Salmon spawn up myriad rivers and streams, some steep and some relatively flat, some large and some small, as well along lake shores.

Along with American Creek, one of those myriad little streams with only two or three hundred yards of fish-able water has also become a favorite. In the fall of 2021 I'm on my third of four trips to this stream. The weather is beautiful. Bright and sunny. Unusually warm for so late in the summer. But stream conditions are not ideal. They are also very different from either of my previous trips. The differences are a good reminder of the importance of the resiliency of the region as a whole thanks the diversity of its many waters, despite bust years on individual streams. On my first visit, conditions were perfect and this stream fished like a dream. Without straying more than a hundred yards from where we started, we caught fish after fish—both rainbow trout and Dolly Varden char—in between thick pods of spawning salmon. On my next visit, a hot dry year especially impacted by brush fires, the water was very low (like so many other Bristol Bay rivers), and while there were some salmon in the stream, and some 'bows and Dollies had followed them up for the egg drop, much of the river was too shallow to hold fish. They were congre-gated in the deep water around the outsides of the wide sweeping bends. Which also meant they were underneath fallen trees. Fishing was slow. We managed to entice a few big trout to take our flies, but only by swinging flies deep beneath logs. The results of having a two-foot-long egg-fattened rainbow trout take a fly underneath a log were predictable. All of us got broken off by at least one big fish that day.

This year it looks (and fishes) like a completely different stream. Winds have limited our choice of tributaries to only this one, and this one is running very high and muddy from recent torrential rains. Crossing where it was too shallow for trout last year now requires one to take their life in their hands. But once we reach a gravel bar where we can fish, we begin to spot rainbow trout almost at once. In this high water, instead of sitting under the outside of the bends where there are raging torrents, several trout sit in the softer water against the inside of the bend. A couple others as well as a few whitefish are visible in deeper water at the tail out. A couple of the trout are real lunkers—as long as many of the sockeye moving around them, and fatter. Rick lands one quickly near the head of the pool. Then another. I land a couple at the tail end. Mine are small by Katmai standards, but Rich hooks into one of the behemoths. It tail-dances in an impressive display of girth and weight. It is easily over two feet in length, and it soon shows that it can pull as well as leap and dance, diving beneath the trees in the deeper water where it wraps the line around a log and breaks it off.

We find several other fish that won't touch any of our flies: flesh, egg, or leech. We try to work upriver. The exceptionally high water makes going tough. There isn't much room to walk the shoreline and we find very few places we can cross. We spot more big trout, but casting for them is hard, and they are picky. Rick hooks another behemoth: a fish well over two feet long. He, also, gets taken under the branches. For several seconds, even while his line is wrapped around a root, the fish remains on the hook. But when he tries to get his line off the branch with a hard tug, he loses it.

I spent thirty minutes drifting flies past one huge rainbow that is two corners up from where Rick hooked his. The trout ignores them all, except on the rare occasion when I get a perfect drift right into its nose. Then it pays enough attention to move out of the way of my fly. Eventually we realize that continuing upriver is impossible; the water has become too swift and deep to fish or cross. We give up and bushwhack down to the original gravel bar. We all find some big trout to cast to. I spend twenty minutes casting to one large stubborn fish. When Rich moves downstream twenty yards, I give up on my fish and cast another thirty minutes to the huge trout Rich had given up on. Another shows up so I can drift my fly past two with the same cast. I am no more successful with two than with one.

The afternoon is winding down. I move back to the upper end of the pool and settle for casting for a smaller pair of rainbows feeding side by side. They prove equally uninterested. I'm watching the two smaller rainbows when a big one moves past me and continues upriver. Assuming it's one of the fish I've already spent half an hour sight-fishing for, I don't plan to follow it. But then Glen calls up to me, and tells me it's a new fish—one he doesn't recognize that looks a bit older and more beat up than the others. I haven't seen those details, but I trust Glen. I have one rod rigged with a flesh fly now, and another rigged with a heavy egg pattern tied with chenille around big brass eyes. It's a heavy fly and gets down quickly without extra weight. The big new trout is only eight yards upriver in some shallow riffle below the gravel where sockeye salmon are dropping eggs. I take a couple steps and cast my heavy dumbbell-eyed egg. It's a good cast and drifts within three inches of the trout's nose. The trout moves. For an instant I can't see the pink of my fly.

Sometimes when sight-fishing I get so eager anticipating a strike that I end up setting the hook before the fish even hits, thus pulling my fly right out of the mouth of a fish. Somehow this time I manage enough patience to wait for the strike. When I lift my rod, I feel the resistance. It feels for a moment like I have hooked the bottom underneath the trout rather than the trout. But then the river explodes.

The battle is hard and tense. The fish is only a few feet from the major snag that has already cost us a couple big fish today. I don't try to fight the fish upstream. I follow it down working hard to pull in toward the beech and keep it out of the logs and roots. Although the fish puts a huge bend in my rod, I somehow manage to keep it out of the roots. I barely keep it out of the second brush jam at the tail of the pool. I am able to land the fish. The biggest of the day and one of the two largest of the week. Certainly the heaviest. For a moment, I'm not thinking at all of bust and boom years for salmon populations. I'm thinking of bust and boom days for trout fishing. With a single fish, this day has just become one of the latter.

6.
Lake Clark National Park as a Classroom (and the Oft-Overlooked Grayling)

It's the last day of July, 2021. The flight from Merrill Field in Anchorage to Port Alsworth is only a few minutes old when our single engine plane flies over the mouth of Susitna River along the northern shoreline of Cook Inlet. Dave looks down out of his window and spots some beluga whales. It's hard to count them; the whales are just small white dots, constantly appearing and disappearing as they alternately surface and dive. Maybe four or five? Dave isn't sure. And then we are past them. Unfortunately, I'm on the wrong side of the plane and I miss the sighting. Cook Inlet belugas are an endangered population. I would have enjoyed seeing them. I'd also have enjoyed seeing seals, which are a common sight at the river mouth. No matter. A short time later, we cross Beluga River where a larger pod of whales is visible. This time I get a look. A few hundred feet below my window, eight oval white blotches sit on a shimmery gray-blue carpet. They are chasing spawning salmon up the river mouths.

The salmon are concentrated enough to draw the whales in, just as they do bears and many other creatures. This is the only time and place I've ever seen the endangered Cook Inlet belugas. In Bristol Bay, however, belugas are more populous and common. They have been known to swim up rivers such as the Naknek and Kvichak. They have even made appearances in Lake Aleknagik twenty miles up the Wood River from saltwater. The salmon that manage to slip through the gauntlet of whales and seals at the river mouths—having already spent multiple years at sea avoiding predation by those same creatures as well as by sea lions, porpoises, sharks, and orcas—will soon provide food for bears, bald eagles, seagulls, trout, char, grayling, and myriad other terrestrial and fluvial creatures in the Susitna River and Beluga River and their myriad tributaries.

Once we reach the far side of Cook Inlet and pass the string of oil rigs extracting resources from below the water surface, our plane turns west into the mountains and then northwest to fly through Lake Clark Pass. Then it's glacier after glacier after glacier on both sides of the plane: small hanging glaciers tucked into alpine hollows above the elevation of the plane; larger glaciers that extend from high above us to far below with fifty-foot *seracs* reaching up to scratch the sky; and frozen rivers of glacier stretching out of sight up long valleys. I'm aware of how important these melting glaciers are to spawning salmon, providing a steady flow of cold water throughout the summer. These slate-gray torrents are flowing beneath the plane. And a quick look up the valleys from which they originate offers dramatic evidence of how quickly the glaciers are disappearing. The visible receding is dramatic, backing long distances up into valleys. It is also geologically recent; no plants have yet colonized the valleys, which are still covered in gray silt.

When our plane turns back in a southwest direction and crosses the peak of the pass, we have entered the Bristol

Bay drainage. We cease following the rivers of melting glacier upstream away from Cook Inlet and start following them downstream, flying over the Tlikakila River toward Lake Clark. Unlike the Susitna and Beluga Rivers, salmon are not yet moving up the Tlikakila. Fed by melting glaciers, and full of glacial flour, the Tlikakila is almost the color of concrete. It has the latest spawning run of sockeye salmon in the Bristol Bay drainage. In 2020, I was able to spend a late September week in Port Alsworth. Since The Farm Lodge had been forced to shut down early that year because of the pandemic, my trip included some fly-out fly fishing excursions with Glen Alsworth Jr., who otherwise would have been busy then. By that time of year, the sockeye run was long over in the various rivers we fished together in the drainage farther south in Katmai National Park and Preserve. Our go-to flies were imitations of rotting salmon flesh rather than egg patterns or my favorite varieties of egg-sucking leech. Up on the Tlikakila, however, the run was just getting started. On my late-September flight back from Port Alsworth to Anchorage, we spotted several brown bears in or along the river. The late salmon run had extended their feasting season, giving them a few more weeks to fatten up for winter.

On this July flight, there are no bears because no salmon have yet entered the river. So instead of looking below me for animals, I scan the steep valley slopes beside and above me for glaciers and waterfalls. On a previous flight through the same pass, our pilot Liz Davidson told me she had tried to count the waterfalls and had given up when she reached one hundred. I don't even try to count. Although it's my eleventh time flying this route in one direction or another, I am not bored with it. The journey is as breathtaking as ever.

I am traveling this time with a dozen others who are all part of an Augustana University class on environmental writing and philosophy: ten students and three instructors, spread out among four planes. David O'Hara, a professor at Augustana and the primary instructor of the class, has invited me and Rob Green to be co-instructors with him. I'm honored and delighted to be a part of the group. Dave is an amazing teacher. He is especially adept at encouraging students to be attentive, ask questions, make connections, and learn the names of flowers. He's also been a close friend for thirty years and we have co-authored three books together (including one on Appalachian brook trout.) Though I only met Rob at the start of the class two weeks earlier, we've already become friends. A former student of Dave's at Augustana, he has gone to graduate school in Montana to study human-grizzly interactions in the Rockies. He is a wonderful writer and photographer, and also the trained backcountry emergency medical person in the group. (The latter of those skills has already come in handy once on the trip.) It is my seventh trip to Port Alsworth, and my third with Dave, but my first with college students. I'm excited by what they will learn, what they will experience, and how it will shape them. I have no doubt the place is still shaping me. Perhaps not as visibly as glaciers and waterfalls have carved the landscape below us, but just as certainly, and far more quickly.

Soon we reach the end of the Tlikakila and a long turquoise lake appears below us, extending out of sight ahead of us to the west. On the map it appears by its English name: Lake Clark. To the Dena'ina people who have dwelt for centuries around its shores and in the surrounding areas, it is known as *Qizhjeh Vena*: a "place where people gather" or the "Place People Gather Lake."[9] Four million acres of Lake Clark National Park and Preserve spread horizon to horizon on both sides of the plane. This national park is where our class will spend its final week. I appreciate that one of the park's stated goal is "preserv[ing] the ancestral homelands of the Dena'ina people"—even if it does use the English name of the lake in the park name. I appreciate also that the park seeks to preserve "an intact ecosystem at the headwaters of the largest sockeye salmon fishery in the world, and a rich cultural wilderness." John Branson, in his book *A 20th-Century Portrait of Lake Clark, Alaska*, notes that the earliest record of human habitation on the shores of Lake Clark dates to roughly 4,000 years ago, but there is evidence in the broader area of human habitation going back at least 10,000 years.

Both the cultural and ecological preservation seem more vital to me than ever. Indeed, one of many important reasons for protecting Bristol Bay fisheries is their importance to the livelihoods of the peoples who live on this land. I'm appreciative of the work of organizations like Bristol

---

9 Dena'ina place names and their meanings, here and following, come from *Dena'ina Elnena: A Celebration: Voices of the Dena'ina*, edited by Karen E. Evanoff.

Bay Heritage Land Trust and their work through the Bristol Bay Fly Fishing and Guide Academy providing opportunities for the native inhabitants of this land to benefit from its economy.

Arctic Grayling:
The Oft-Overlooked Character

The plane descends lower over Lake Clark. We get closer to that beautiful turquoise gem. Soon our plane turns and we fly over the Tanalian River. At least this river, like Lake Kontrashibuna out of which it flows, still has the vestiges (Anglicized though they are) of its description in the Dena'ina language: the *Tanilen Vetnu* which translates as "flows into water stream".

When the four planes land on the dirt airstrip in Port Alsworth, and our gear gets settled into our cabins at The Farm Lodge, it's time to experience some of the ecology, history, landscape, and culture of the town, of Lake Clark National Park, and of the Bristol Bay drainage. As a teacher who happens also to be an angler, the first thing that pops to my mind is to look for native fish in their natural habitat. So I'm thrilled when the class decides to explore down the shoreline of *Qizhjeh Vena* to the mouth of *Tanilen Vetnu,* the stream flowing into the greater water. I pull on my waders and wading shoes, grab my fly rod and some flies, and I'm ready. Rob also grabs his gear, as does Josh, a graduate student taking the class for credit. A few days earlier, I'd watched Josh delight in catching his first Dolly Varden char, enticing it with a dry fly in a small glacial river running through an alpine meadow in Chugach State park. He's now eager to see an Arctic grayling. And though Rob is a proficient fly fisher with experience in Montana, he is still waiting to catch his first Alaskan fish of any kind. My own plan is to help Kira, one of the youngest students in the class, catch her first fish of any kind on a fly. I'm looking forward to seeing her expressions of delight and wonder as she holds a grayling in the water for the first time. I want *all* the students to appreciate the subtle beauty of a grayling even while they soak in the majestic landscape that surrounds us.

It took me some time to come to deeply appreciate the Arctic grayling. Like trout, they are members of the cold-water *Salmonidae* family. Their full scientific species name, *Thymallus arcticus*, was recorded in 1776, but the genus *Thymallus* goes back another eighteen years to when it was first applied to their cousins the European grayling, *Thymallus thymallus* because they smelled vaguely like the herb thyme. When I first encountered grayling twenty years earlier on a tributary of the Nushagak, they were so plentiful and hungry that I struggled to keep them off my line while I was trying to fish for rainbow trout and Arctic char. And because they were not the fish I was trying to catch, and had on first glance a somewhat more drab coloration than species of char, I failed to pay close attention to them, and thus also failed to notice how beautiful they really area.

Cecilia Kleinkauf's excellent book *Fly-Fishing for Alaska's Arctic Grayling: Sailfish of the North* provides a short but excellent account of the life history of a typical Arctic grayling. Grayling emerge from their eggs—about half the size of trout eggs—after two to three weeks. When they hatch, they can barely swim. They remain in backwaters with large schools of fry, feeding on zooplankton, until they reach about two inches in length. Then growing one to three inches a year, it takes them five to seven years to reach reproductive maturity and lengths of about a foot long. In some locations with good conditions, they can live over thirty years and grow to be well over twenty inches.

Grayling are also great travelers. Not only will they travel long distances with the trout, char, and bears to follow the sockeye egg drop as Schinder pointed out, but Kleinkauf also notes that they can travel up to a hundred miles to spawn—usually right around spring iceout—since good quality spawning gravel is very different water than the deeper pools where they overwinter which also may be different from good feeding areas. And while large grayling are omnivorous predators that will even eat rodents as well as other fish, small grayling are an important food source for many other creatures. Pike and trout consume them as do eagles and osprey and members of the weasel family including otters and mink.

Their most striking feature at first glance is what gives Kleinkauf the subtitle of her book: they have dorsal fins that are very large for the size of their body, making them look not unlike a smaller freshwater version of a sailfish. Those fins also make them very good fighters for their size. And though I haven't quite broken that twenty-inch mark with a grayling, I've come close and learned that a nineteen-inch grayling, especially in a strong river current, can break a line or bend a rod. What I've come to appreciate just as much as their unique silhouette and good fighting ability is just how beautiful they are—and how surprisingly colorful they are, especially when seen close in the bright light of a long Alaskan summer day. What from a distance appears a drab olive is a mosaic of numerous brightly colored scales in a range of shades, including some black scales. In the males, the spines on their dorsal fins are beautiful enough to inspire poetry.

For all of these reasons, I am eager to introduce them to others in the class, who can appreciate them with me. When we arrive at the river mouth, I suggest that Rob and Josh work the deep soft water of the eddy behind a point of land upriver on the near bank. I know from experience that it's a good honey hole, and there is plenty of space for both of them to cast. Kira, however, is limited by her hip boots and lack of casting experience. Without wading deeper than the height of her boots, she won't be able to get a fly to where the grayling sit along the seam. So I lead her further downriver where a shallower gravel bar follows the edge of the current. The water is just below the top of her boots, so we move slowly.

It doesn't take Rob long to make a connection with his first grayling. I'm forty yards downstream, but I momentarily leave Kira and wade back against the current with my net to make sure he can land the fish and get a clean photo. I haven't even gotten back to Kira before Rob lands his second and third fish. Josh is soon catching fish also. I ignore them now and focus on helping Kira. We've already tried a generic attractor nymph, which is usually successful in this spot. However, the current is swift and Kira doesn't have the hang of mending line to control the drag, or of sensing a fish take a fly below the surface. Then I notice fish feeding on the surface a little farther downriver where the gravel ledge drops into the lake. One large grayling is repeatedly slapping something, not much farther away than twice the length of my fly rod. I see some small caddisflies hatching, so I switch to a black elk hair caddis and help Kira cast to fish twenty yards downstream. With a dry fly visible on the surface, it's easier for her to see and manage the line drag, and she can use the length of the rod to get at least a dozen feet of drift without having to mend the line. It takes her some time to get used to seeing the fly on the water, seeing the fish hit the fly, and setting the hook. Eventually, with a little help, she manages to get one on. She has a large smile as she plays her first fish alone. We keep the fish in the net under water and walk it a few steps closer to shore. There she briefly and carefully lifts it out of the water as photos are snapped from shore and her fellow students cheer her. Then she completes the release. We head back to our spot. Eventually it all comes together for her, and Kira successfully hooks two more fish without help.

I think back on a passage from the wonderful book *For the Love or Rivers: A Scientist's Journey* by Kurt Fausch, a fisheries biologist, stream ecologist, and cutthroat-trout expert who taught for many years at Colorado State University. It's an important book about stream ecology and about how we learn and care about the world. It's also the book that provided my first lesson on Dolly Varden char from Fausch's research in Japan. I have quoted the book more than once. "People learn best when information enters through their peripheral vision, out of the corner of their eye. . . through stories," Fausch writes. He adds, "Native trout can return to their streams only when people view their existence as essential to their lives. And this can only happen if we see and touch these amazing creatures for ourselves."

Fausch was focused mostly on Colorado's native cutthroat trout when he wrote that, and about places where those fish were threatened or had already been extirpated due to human influences such as water removal, development, resource extraction, and invasive species. Better even than restoring native fish, though, would be to preserve them in the first place: avoiding the mistakes made in so many other places rather than trying to repair damage after it's too late. To that end, I hope the experience of seeing the amazing creature known as a grayling will help the students in the class view the existence of those fish—and the existence of the salmon that will also soon swim in these waters, and the protection of the places where those fish live—as essential to their own lives.

Sadly, the ongoing pandemic means that the visitor center at Lake Clark National Park is closed. Only the outdoor displays are open. The class visits these the next morning. The historic cache for storing dried salmon is a good reminder that salmon are not only vital to the ecology of the region and to creatures such as bears and eagles. They also remain an important part of the culture and livelihood of the Dena'Ina peoples who have lived on and around Lake Clark for centuries, and to the Yupik peoples whose ancestral land covered much of Bristol Bay drainage further south and closer to the coast.

After visiting the displays, we hike up the hill to Tanalian Falls, a couple miles upriver from the river mouth where we visited when we first arrived. The misting, thundering thirty-foot drop creates an awe-inspiring (and photo-worthy) natural feature, and a great place for a picnic lunch and class discussion. It also marks a significant ecological divider. As great as they are at leaping, salmon cannot climb *this* falls. The uppermost stretch of Tanalian River above the falls, as well as Lake Kontrashibuna and all its feeder streams, are devoid of the MDN that come in salmon eggs and decaying or digested salmon flesh. The nutrients that feed the whales and bears are simply not available here. Above the falls, the large predatory fish are lake trout and Arctic char whose diet includes insects, sculpin, sticklebacks, and in some cases each other. As numerous as the grayling are below the falls, there are no grayling above. They had not colonized the water before the forces of glacier and erosion had created the impassible barrier.

There is a small run of sockeye salmon that does move up the river later in the summer. Although one can occasionally see a bright red shape making futile attempts to leap the falls, most sockeye don't even make it that far. They spawn farther downriver or on the gravel shore at the river mouth where last night I helped Kira catch her first grayling. So even here at the falls, the ecosystem is not dominated by the egg drop as it is in so many rivers in Katmai National Park or on the Nushagak drainage. You will not see brown bears wading these falls in hopes of catching salmon as they do at the more famous Brooks Falls in Katmai. Yet their eggs are so appealing that even the small number of salmon that make it this far can entice some trout from Lake Clark to chase them upriver in late summer.

I have also been told by National Park Service biologists and technicians that they have spotted big lake trout in the pools below the falls. They aren't sure where the lake trout came from. It's possible that some wash down over the falls from Kontrashibuna, and have no way to return. But they might also have chased spawning sockeye salmon three miles upriver from Lake Clark. Although lake trout are a fully lacustrine species of char that spawn only in lakes, and are not known as river-dwellers, they will enter rivers either to migrate (in search of new habitat) or to chase food. I like the later theory, though I have no evidence to support it. Although I don't really expect lake trout or rainbow trout in the water today since there are no sockeye salmon present, they still reside in the back of my mind as I ponder what fly to use. Branden Hummel, who teaches science at the Tanalian School in town in addition to his job guiding with the Farm Lodge in the summers, is the one from whom I first heard the term "unicorn" used to describe a rarely encountered fish—or more accurately a fish encountered in a place it isn't usually encountered: a rare creature you believe might be present, but rarely (or never) get to see. I believe in these unicorns. Two years earlier, after seeing sockeye trying to leap the falls, I began fishing below with an egg-sucking leech and surprised myself by catching a rainbow trout amid the grayling. In many hours spent over many years fishing the Tanalian, it is still the only one I have caught in the river. Maybe now I can hook the unicorn lake trout.

After eating streamside lunch with the rest of the class, I rig up my fly rod and change into sandals. Today is the earliest in the year I have visited this spot, and it's the highest I've ever seen the river. Water roars over the falls, white and lime green, foaming high in two turbulent cauldrons. The plunge pool on the near side is usually my favorite place to cast for big grayling. Today it is unfishable. I see a few fish rising in the slack water, but they are small. I'm going to need something heavy and deep to get into one of the bigger fish. I tie on a small but heavy olive bead-head wooly bugger and start drifting it along the edge of the foaming water down into the deep soft water above a big boulder. On my first drift in front of the boulder, a dark shadow comes up from the bottom, swirls my fly, and misses. Three or four casts later I land a fat, hard-fighting grayling bigger than most I have caught here. I hook two more and land one of them before they stop biting. I try seven more flies: a streamer, two other shades of wooly bugger, and four different nymphs. Almost all elicit one flashing look from the mysterious dark shadow deep in the pool. I feel it hit the hook once. But we never connect.

I tell myself that the dark shadow is just a large grayling and not a unicorn. Eventually I give up trying to prove myself right or wrong. I sit on the shore and chat with students and eat a snack and take pictures. Before we leave, however, I wade out one more time before changing from wading sandals back to hiking shoes. I tie on the one color of wooly bugger in my box that I haven't yet tried. As in the past, the big shadow rises from the depth. Unlike before, this time it takes my fly hard and I manage to make a firm connection. A few minutes later I land a large fat lake trout, with a bright orange belly. I'm not at all sure what its presence there tells me, except perhaps that unicorns exist. Keeping it in the water, I bring it a few short steps to shore and let the students—who have already seen grayling and Dolly Varden in the first two and half weeks of the class—admire a new native Alaskan fish species.

Part III: Education, Economies, and an Epilogue

7.
Casting Flies and Asking Questions

I am still in the Bristol Bay drainage with Dave and his Augustana University Class. After a short stay at the Farm Lodge in Port Alsworth, we take a wet and choppy boat ride up Lake Clark with all our gear to spend several nights at National Park Service historic Joe Thompson Cabin. The cabin is set several strides back into the woods near the shore of a semi-sheltered cove on Lake Clark. Trailing red currants line the shoreline. It's the start of August and they are at the peak of their ripeness. From the back side of the cabin, a hiking trail leads up the ridge past patches of watermelon berries and on up to alpine meadows of wild blueberries. Little forage fish cruise the shoreline all day, and when the water is calm larger fish occasionally rise off the shore beyond casting distance.

In addition to exploring this new part of Lake Clark and the Bristol Bay drainage, Dave has also arranged a special classroom for a day: pilots and guides from the Farm Lodge are going to take the class by floatplane down to Katmai National Park and Preserve for a bear-viewing excursion. In many ways, it is the culmination of the class experiences for our students.

It's shouldn't be a bad trip for me, either.

Baby Brown Bears, and Big, Line-Breaking 'Bows

It's been three days since our grayling expeditions to the outlet of the Tanalian River where Kira's caught her first fish on a fly, Rob netted his first (of several) Alaskan fish on a fly, and Josh managed to land and release his first Arctic grayling. Two days have passed since I caught the unicorn lake trout at Tanalian Falls. Yesterday, Rob and Dave went on the bear-viewing trip with half of the class while I and the other half hung out at the Joe Thompson cabin and workshopped some of our writing from the previous week.

Today, I join the other half of the class on an expedition to Katmai National Park and Preserve. Based on the enthusiastic reports from the first group (mingled with a few complaints about the early morning cold), the second group is eagerly anticipating the day. I am too, in part because I will have my fly rod with me with reasonable hopes of using it. But first and foremost, it is a bear-viewing trip and one of our final "classrooms" of the summer. The students will not only see bears gorging on spawning sockeye salmon more than a thousand feet above sea level and a hundred river miles from salt water. They will also witness the process of marine-derived nutrients impacting the entire ecosystem through adept work of the bears (and other creatures) spreading those nutrients across the landscape in uneaten carcasses and excrement. We will have to be careful all day to avoid stepping in these piles of nutrients left behind by the bears—piles of nutrients that work their way into the soil and feed the roots of plants rather than working their way into our boots to be left behind in airplanes and tents.

We have snacks for lunch, along with water and lots of layers of clothing for a forecast of a cold morning turning into a hot afternoon. My friend Branden Hummel is our guide. I ride down in a plane with Glen and continue to learn about some of the lakes and rivers we see below us. We first fly over fog-covered lakes and deep valleys in the southern part of Lake Clark National Park. Once we reach Lake Iliamna, we're past the fog. The view clears in every direction. We catch glimpses of Mount Iliamna and the Alaska range of volcanic peaks to the east, and see freshwater seals hauled out on an island below us. South of Lake Iliamna, as we come to the northern portions of Katmai National Park and Preserve, we start to see bears in the rivers below.

Soon we are exiting the float planes and marching across the tundra. I'm excited for the students to see the bears up close. I think it will be an unforgettable life-changing experience for them. But I'm also aware of how good the fishing is. It is good for the same reason bear viewing is good: the river is loaded with spawning sockeye salmon right now. Students were almost as amazed by the massive red schools of salmon visible from the airplanes as they were by the bears. And the egg drop has begun. The bears are gorging on salmon. And Dolly Varden char and rainbow trout have also moved up out of the lakes into the rivers to feast on drifting eggs. I've had some of my most memorable fly-fishing experiences casting for trout and Dolly Varden in this spot. My feelings are mixed. I want to be with the students seeing the bear. I'd also like a chance to take at least a few casts.

When we get across a stretch of tundra and start walking just a dozen yards from the riverbank, I see my chance. Straying only a few strides away from the group, I can get in a quick cast or two. I tie on my own egg-sucking leech pattern: big tungsten dumbbell eyes wrapped with pink chenille to form the egg, and below that a black chenille body wrapped with black hackle and red wire in the style of a wooly bugger including a long black marabou tail. Keeping half an eye on the others as they slowly move away, I propel my heavy fly forty feet out and downstream onto the inside bend on the other side of the current. As my fly gets out of the soft water and hits the seam, a huge rainbow trout takes it. The fish tears off downstream and my reel starts spinning and buzzing like a teenager's wheels on wet pavement. Having broken off a lot of big fish on this river before, I'm nervous about putting too much pressure on the line. In one blink of an eye the fish has me to my backing, and in another blink it leaps high out of the water and spits the hook.

First cast, and I've already lost a big fish. At least I got a glimpse of what I lost. And I kept my fly.

I take a deep breath. No way I can quit casting after that. The students are with Branden, who is also carrying the bear protection on his hip. They are safer than I am. I continue fishing my way down the river, moving vaguely parallel with the group but keeping only a tenth of an eye on them. Two corners past my first fish, I hook a second. It's another huge fish. This time I make sure of my hook set, and I put more pressure on the line to avoid another hook spitting. But even with increased pressure, I can't turn the fish in the current. The river is too deep and swift on the near side to wade after the fish, so I follow it from shore, rod lifted over the taller bushes as I look for slack water where I might be able to step into the river and bring the fish in to net. I'm moving farther from the group that is now walking perpendicularly to me and growing smaller in the distance. I want to land the fish quickly now so that I can keep up with them. I'd also rather not surprise a bear napping in the bushes after gorging on salmon eggs. So rather than traipsing through brush working around one more point in search of calmer water, I get to a little eddy and try to force the fish in there. Mistake. The trout breaks my 2x tippet and this time I lose both the fish and the fly.

That's it for fishing for much of the day. I hustle to catch up with the students. And I have a truly enjoyable time watching bears chase salmon in the river below us, and watching students watch bears chasing salmon. The delight and awe in their expressions is palpable. We spot a sow with yearling cubs napping on the opposite shore. A big blond male walks right below us. An even bigger blond works the river downstream of us, with another darker bear occasionally appearing from around the next bend. For the most part, the bears ignore us. As has been the case in previous visits to spots like this around Bristol Bay, we observe dynamics among the individual bears including a pecking order of which bears get the prime spots. We witness a bluff charge as a sow with a cub drives off a young male. Yet there is so much food in the river right now that all the bears seem able to get as much as they want even if they don't have the prime spots. They only need to hunt for a minute or two and they have their catch and climb up on the bank to gorge.

The temperature has warmed up quite a bit by midday. The extra layers are peeling off. When there are no bears visible below, we decide to drop down off the bluff into a little inlet stream—a snow-filled gully flowing in from the north—in order to see the salmon from stream level. But when we step up to the edge of the bluff, we see right below us a sow bear and three cubs on the snow. Like us, they seem to have gotten too warm. They can't peel off extra layers, however, so they've come up onto this large patch of snow. The sow starts looking at us intently and yawns in a way that (as Branden explains) indicates discomfort with our sudden appearance and nearness to her and her cubs. We back up along the bluff another twenty-five yards to give the quartet more space. The mother bear seems satisfied with the increased distance, or with the fact that we yielded ground. She stops eying us, and lies down on the snow.

And then comes the highlight of the day. For the next hour we watch the four bears napping and wrestling and staying cool on snow. The family dynamic is wonderful entertainment. One of the little cub brothers (or maybe a sister) keeps biting his siblings' ears and constantly wanting to wrestle and play. The others are more interested in the napping. Finally, when the mother has had her nap—or perhaps when they've all gotten tired of the annoying sibling's antics—they head back to river for more feasting. Mom does a full-on plunge into deep water and catches a salmon. The annoying cub plays tug of war with its mom to get the salmon. The mom plays along for a moment and then yields the fish. Another cub hunting a shall gravel bar manages to get a salmon on its own.

Our hearts are full of joy.

And yet…

And yet we have questions also.

A Different Economy

Just a week earlier, our class was camping at Right Beach in Kachemak Bay State Park near Homer, about ninety miles east of the Alaska Range and the boundary of the Bristol Bay drainage. We'd had discussions about what *wilderness* is, and whether the place we were camped could or should be designated as such. We were in a roadless state park spanning 400,000 acres. Thick forests and rugged inhospitable peaks stretched for many miles to our south and east. The thirteen-mile-long Grewingk Glacier was less than a day's hike from our tents. Black bears were common in the area, along with moose. Harbor seals hunted just yards off the beach in front of us. Sea otters swam by our cove in the mornings and evenings. We often had as many as five bald eagles perched atop the bluffs or tall trees around the cove calling to one another over our heads with such expressive and almost intimate vocalizations that I felt like I was in their living room—or, more accurately, their dining room; on multiple occasions we saw one or another of them catch a fish nearby and consume it on the beach in front of us or up on their perches. When we went to the spring for drinking water, we had to bring bear spray, and navigate through a thick patch of devil's club. The only human structures we could reach within a day's hike were a couple yurts and public-use park cabins. All of this made the place *feel* wild in some ways. And when news came to us via our cell phones of a powerful earthquake off the Alaskan coast, with a corresponding warning of a possible tsunami, we had no possibility of rescue or evacuation. Civilization was too far away. We were on our own. That really made our setting seem remote and wild.

Except that the tsunami warning that had reminded us of our remoteness and vulnerability came to us *on cell phones* that had three or four bars of service. All the while our class camped there, we could look across Kachemak Bay toward Homer and see one or two big freighters or oil tankers anchored in the bay, close enough to hear the low thrum of their powerful engines. Both sport fishing and commercial boats worked the bay, the shoreline, and the cove. One boat spent a whole afternoon doing three sets of their huge purse-seine right off our beach, with the skiff close enough that we could stand on the beach and converse with the skiff pilot. At night, we could see the lights of Homer glittering off the water. When the tsunami warning came, we saw the flashing blue lights of police vehicles evacuating Homer Spit. The cliffside homes of Halibut Cove, though not reachable by hiking except via a long circuitous route, were close enough that we could see individual windows and doors. It was another reminder of both affluence and influence: the affluence of a community that can afford such extravagance in the midst of a remote setting, and the human influence on this remote bay. Two big bunches of buoys marked an oyster farming operation between us and the homes of Halibut Cove. It was hard to think of ourselves as being in wilderness.

I have similar questions here on our bear-viewing excursion, at the edge of a picturesque bluff on a windswept tundra in the "wilds" of Katmai National Park and Preserve, with at least ten brown bears within a mile and a half of us, and several thousand salmon swimming upriver through the bottom of the canyon below us. Our cell phones have no bars here. John Branson, who at the time of our outing is in his final year as historian for Lake Clark National Park, made a presentation to our class and pointed out that fifty thousand square miles in the Bristol Bay area contains an estimated eight thousand to ten thousand brown bears and only seven thousand humans. I am surrounded by spectacular beauty: rocky ridges whose bare gray rock faces are etched with snow and lichen, wildflowers and berries strewn across the tundra, the carved tan of the canyon and bluffs at my feet, and the bold reds, oranges, and dark greens (and ostentatious humps and kypes) of spawning sockeye salmon whose impending death only seems to bring about their most ostentatious displays. I am delighted even by the variety of fur colorations of the brown bears. I've been awed in the past by magnificent animals in zoos, but it is a quite different and more powerful experience to see these creatures in the wild. I have no doubt that the sense of wildness, remoteness, and isolation makes the experience more delight-filled for me. The setting brings me to the point of tears, knowing that such places as this exist in the world. I suppose that's good, especially when it moves me to work to protect those places.

But I also know that, like our setting at Right Beach a week earlier, some of that sense of wildness is illusion. The bluff that we walk along not only has the trails of the many bears that wander the area leaving piles of nutrients on the soil, but also the more worn-down trails of human passers-by: numerous bear viewers and fish catchers like ourselves. Our trail has left a scar on the thin soil of the fragile tundra. Late in the morning, a party of at least twelve persons walked along the south side of the bluff across from us, pausing when they came upon some bears to take out huge cameras and snap photos for hours on end, just as we are doing from the north side of the bluff. At least ten times over the course of the day, a float plane lands or takes off nearby. And not far away is another lake where the float-plane traffic is even more intense. On some days, the lake looks like a parking lot for planes. The ways that we all have found delight in the beauty of this place has a cost.

And even as I ponder the impacts of human habitation, I recognize a strain of euro-centric arrogance running through my veins. I hear people speak of "human habitation" of these lands as a recent phenomenon, dating to the arrival of Russians and then later Europeans. But a variety of peoples have been living in Bristol Bay for millennia. John Branson told us of fifteen longhouses that had been built centuries ago near Kijik, across the lake from Port Alsworth. One old settlement on Hardenberg Bay dates back 3700 years. Russians and Europe's did not discover or settle the area. They invaded it. But I do believe that the presence of us invaders has had a far more intense impact on the land and ecology of Bristol Bay than the Dena'ina and Yup'ik peoples who were here before.

As I think about our class trip to Kachemak Bay, one moment stands out in my memory. A commercial fishing boat is doing a set of its purse seine in the cove in front of us. While the larger main vessel is a couple hundred yards offshore unrolling the giant net, one of the crew is running the smaller skiff attached to the end of the net. She is keeping the skiff and net as tight as possible to the shore so that salmon don't escape around it. She is so close, I could reach out and touch her boat from the shore with the nine-foot saltwater fly rod I had just used to land a salmon from the beach. So I and several members of the class are standing on the beach shouting over the roar of several hundred horsepower of revving twin outboard engines as we converse with the skiff driver. Her name is Ursula. She tells us that the previous day's by-catch included a three-hundred pound halibut. At thirty dollars or more per pound, the final market value of such a fish could be close to ten thousand dollars. She says the halibut was a female at the prime of its reproductive maturity, and notes that the future of halibut in this area is in fish like that one. Then she adds, with a hint of regret, "If I were a millionaire, I would have let it go."

But she isn't a millionaire. She has bills to pay. Rent. Food. Insurance. Gas. And so they kept the halibut. It will fetch them only a fifth or so of its final market value, and yet it still represented a good chunk of cash for the crew: a few hundred dollars apiece. Our class was then left discussing what sort of economy might have prompted her to release that fish instead of keeping it. What would such an economy look it? How can we change our very definition of "economy" to include the sorts of cultural choices that would make it worthwhile for commercial salmon fishers who find ten thousand dollars' worth of halibut in their purse seine to let that halibut return to the oceans and breed?

I think of Tim Troll and *Bristol Bay Heritage Land Trust* and their work to create such an economy in the Bristol Bay drainage, helping to buy conservation easements on the land through which a Pebble Mine haul road was tentatively planned—easements that would make it economically more viable for the native peoples of the Pedro Bay area to protect the land, the salmon, and their historic lifestyles than to sell it off for the mine development. I think also of the fact that many anglers will spend over $2000 a day—and in some cases a lot more than that—to fish some of the more famous rivers of Bristol Bay. When John Branson put me in touch with Tim Troll a few years earlier, I learned about the *Bristol Bay Fly Fishing and Guide Academy* which Troll helped launch back in 2008. In many ways, it's an effort to change the economy: training local youth from the area, especially native youth, to be fishing guides so that more of that wealth flowing into the region through the hands of fly fishers and sportspeople would end up in the hands of those who lived in the area, grew up in the area, and had the most to be gained or lost based on how that land got used in the future. The Academy has not only helped improve the opportunities and finances of numerous youth over the decade and a half since its founding; it has also helped shape an economy, and the people who live in and with that economy, so that it motivates conservation and long-term ecological health rather than short-term profits through resource extraction at the cost of the future.

I think also of the global demand for metals and power—especially electrical power— that drives the economies of resource extraction and hydroelectric dam construction. I think of my own part in that economy. Every time I buy a new computer or update to the latest cell phone, I am driving that demand for copper and lithium batteries and increasing the likelihood of a Pebble Mine in Bristol Bay or somewhere else. My use of social media to share my fishing photos and videos increases the demand for big data facilities, which require not only heavy metals to build the computing equipment but also massive amounts of electricity to run the data farms after they are built. In the summer of 2022, I learn about a possible hydroelectric dam project on a Nushagak tributary with a location currently inside the protected boundaries of Wood-Tikchik State Park. The project, if approved, would require the park boundaries to shrink in order to dam up some of the waters flowing out of the park and down into Bristol Bay. The project would supply a local native village and not a big data center. I understand that need for power. Yet it would still be a dam impacting some of that important and diverse waters of the Nushagak drainage. I think even of the fact that this largest state park in the country has only a single employee and a budget of less than fifteen thousand dollars a year. Our current economic system has professional athletes who make about ten times that much *for a single game or single week.* What might a different economy look like in which we valued rivers, conservation, soils, and native cultures as much as we do being passively entertained?

My employer tells me that my laptop is now three years old, and I'm eligible for a replacement. I can get a newer computer with the latest chips and faster speeds. After a couple days of toying with the temptation, I let them know that I'm fine keeping my current computer for another year or two.

Later in the afternoon, after our day of bear-viewing, the planes return to Katmai National Park to fly us back to the Joe Thompson Cabin. Branden lets me know that in the given wind conditions, there is a limitations on weight for take-off. They will have to make two round trips to get everybody back to Lake Clark. Most of the students, though they have had a wonderful day, are tired and more than ready to return. But Glen and Branden both know me well enough to know without asking that I will be happy to stay behind and return on the second trip. Eager. Ecstatic, even. It means another hour and a half to cast flies to big trout on one of my favorite fishing spots in the world. Josh volunteers to stay with me so that I won't be alone.

Almost at once, I break off another big rainbow trout on 2x tippet. Another precious egg-sucking leech is gone, but it's the lost trout that hurts the most. I replace the tippet. I land my first rainbow. Then another. I land two fair-hooked sockeye salmon, and a third that gets hooked in the fin. A fourth, the largest sockeye of the day, hits my fly aggressively and manages to break tippet just as I get it to shore, and I lose a third egg-sucking leech. But this one doesn't hurt as badly. It's been a wonderful and unexpected gift of ninety minutes.

8.
A Thin Blue Line[10]

I walk up the gravel shoreline, trying to avoid mounds of fresh bear dung underfoot while still focusing my gaze on the water, looking behind the gaudy red and green forms of spawning sockeye salmon for a more subtle movement: a mere shadow on the gravelly streambed. I am trying to pick out a rainbow trout or Dolly Varden char sucking up the eggs drifting down past them, with the hope of getting one or two sight-casts before the fish sees *my* shadow, mistakes me perhaps for a hungry bear, and slinks off into the deep bend.

The river I am fishing has no name. None that appears on a map, or that my guide and friend Glen Alsworth Jr. has ever heard. It's just a mysterious thin blue line in the vastness of Katmai National Park and Preserve—visible only if you know where to zoom your map or satellite image in to one shoreline of one lake, and look closely at just the right spot. Yet when I speak with Glen about that river, I do so in hushed and reverent tones, referring to it as "the unnamed river". He knows exactly which stream I'm speaking of.

As I think about this unnamed stream, which I do quite often, I ponder an essay by Aldo Leopold, "The River of the Mother of God", published nearly a century ago (1924). Referring to another mysterious river, Leopold writes, "All

---

10 An earlier version of this chapter appeared in *The Drake* as "A Thin Blue Line: Mystery of an Unnamed River". *The Drake,* Winter 2020-2021, 104-106.

that I remember is that long ago a Spanish Captain, wandering in some far Andean height, sent back word that he had found where a mighty river falls into the trackless Amazonian forest, and disappears. He had named it *el Rio Madre de Dios*. . . . [E]ver since some maps of South America have shown a short heavy line running eastward beyond the Andes, a river without beginning and without end, and labelled it the River of the Mother of God."

My Alaskan river shares little in common with Leopold's South American one. Far from being a "mighty river", it is just a small stream: a thin blue line not a thick black one. In most places, trees from both sides touch overhead, more like a little Vermont brook trout stream than a mighty Alaska salmon river. With the exception of the final few hundred yards before it reaches its terminus at the lake, my unnamed stream is too small even to fish.

On my first visit there with Glen, and with my close friend David and his son Michael, the four of us made it only a hundred yards up the stream before we came upon a big sweeping bend with an undercut bank, dozens of sockeye salmon in the shallows, and enough big rainbows stretched out behind them to keep four of us busy for a day. I explored a hundred yards upstream on my own just to see what was there, but I didn't have to do that for the fishing. I remember landing seven fish that day, the largest a 27" rainbow trout that had been holding at the bottom of two feet of swift water behind a sockeye redd: an almost invisible shadow in the blur of the riffles and dappled light. I also remember the even bigger trout at the tail of the pool, that all of us took turns casting to. That one was the happily egg-and-flesh-stuffed, and very selective trout. The one none of us hooked.

On my second visit with Glen—accompanied this time by my nephew Michael and my friend Rich—when a statewide drought and heatwave had resulted in very low water (and fires across the state), which in turn resulted in skittish trout hiding under fallen trees and overhanging branches, we worked our way upstream about as far as we could go looking for trout willing to come out from under the logs. A big brown bear standing midstream eventually turned us around, but the water had become so shallow and the stream bed so narrow above the previous little tributary stream that we couldn't have gone much further anyway.

We didn't turn around without some good trout photos, though. Spawning and rotting sockeye salmon were thick in the stream (hence the bear), and the trout had a veritable feast of eggs and flesh floating past them. Most of the fish that I made a brief acquaintance with, I enticed by drifting a big black marabou egg-sucking leech under the edge of one of those fallen trees. It doesn't take a two-foot long rainbow or Dolly more than a wink of an eye to grab a fly, and bring it two feet right back into its nest of trees. As Michael, Rich, and I all discovered on multiple occasions, even a 2x tippet is no match for a log, nor can a nine-foot five-weight fly rod prevent a five or six-pound rainbow from taking a fly a few feet downstream (or farther if it wants).

But in addition to many flies left embedded in wood, I'd landed a few fish—mostly rainbows, but also one gorgeous Dolly with magenta pearls—out of a few deeper riffles. Glen, who preferred a flesh pattern to an egg-sucking leech, also brought a few bows briefly to his net, including one nearly 28" long and so girthy that its fat oozed over the sides of Glen's fingers as he held it at the surface of the water for a photo. He'd spotted it deep under a fallen tree, said aloud it would never come out, and cast to it without much hope. When his pink fly fish hit the water, that rainbow bolted six feet out from the tree and slammed it more like an ambushing northern pike than an egg-sucking trout while I was landing its smaller cousin fifteen yards downstream.

Yet for all the delight I had in finding, hooking, and sometimes landing those fish, it is the sense of mystery and awe of that unnamed river that keeps my thoughts going back to it: the fact that the river has no name, and barely shows up on a map, many miles from the nearest road or even the nearest dirt airstrip or wilderness lodge.

Later in Leopold's essay, he writes more about his *el Rio Madre de Dios* and what it meant to him. "That short heavy line flung down upon the blank vastness of tropical wilderness has always seemed the perfect symbol of the Unknown Places of the earth" which have been reduced "one by one, until now there are none left." He laments, "the time is not far off when there will no more be a short line on the map, without beginning and without end, no mighty river to fall from far Andean heights into the Amazonian wilderness, and disappear. Motor boats will sputter through those trackless forests, the clank of steam hoists will be heard in the Mountain of the Sun, and there will be phonographs and chewing gum upon the River of the Mother of God. . . . speaking geographically, the end of the Unknown is at hand."

Even while I was trying to lure trout with an egg-sucking leech on an unnamed river, not far away (at least by Alaskan standards of distance), a huge mining company was working on the roads and infrastructure that will enable (if ultimately approved) massive amounts of toxic metals to be hauled across the landscape from the proposed Pebble Mine. My unnamed river is not directly threatened by the mine or the haul road—though many other Alaskan rivers are, along with the health of the entire Bristol Bay watershed into which those rivers flow, and which is responsible for about half of the world's annual sockeye salmon catch. But then I think of the dramatic increase in "traffic" in the area when a multi-billion dollar mine goes into operation, and all of the other collateral damage. I wonder how long a river can last without being given a name, without it's mystery and anonymity being taken away. That Leopold's essay was in part a lament of the road-building efforts across the continent that he felt was destroying the continent's wilderness makes his essay more relevant—makes the connection more apparent between his *River of the Mother of God* and my unnamed river.

My son recently told me that it's possible to identify unnamed geographic features and give them official names. I am not tempted. Even if the unnamed river had a name, I wouldn't use it. I hope it never gets one. I'm thinking about it now. About that big mysterious rainbow trout that none of us could entice.

Epilogue:
A Fool Standing in the Rain

As this book began with a tale of catching a Dolly Varden outside of Bristol Bay, I suppose it's appropriate to end with an account of being inside the Bristol Bay drainage and *not* catching a Dolly.

In 1970, about two years before my first wilderness fishing trip when I was introduced to *Salvelinus fontinalis* and *Salvelinus namaycush*, Van Morrison released his album *Moondance*. The opening track of that album—the song preceding the famous title track—is the less well-known song "And It Stoned Me". In the first verse of this song, the narrator is on his way to the county fair with Billy. The narrator has his fishing rod and tackle, and so listeners to the song can only assume they are going fishing. That's when the rain starts pouring down.

Eventually, when the sun comes back out and they dry off, the narrator and Billy abandon their fishing rods in favor of a swim. Being a song (at least initially) about two friends (or perhaps siblings) going fishing, it has always been one of my favorite Van Morrison tunes. It's Van Morrison, so I'm picturing some rural village in Ireland. I imagine they were fishing for brown trout, though Ireland also has some famous Atlantic salmon fishing, and even a few lakes with Arctic char. Indeed, Old Irish, also called *Gaelic*, is the Celtic language that the word *cera* and the name *char* have often been associated with.

One of the joys of writing this book was meeting and getting to know Glen Alsworth Jr. In addition to running his family business, and helping me create the itinerary for most of my trips to the Bristol Bay area, he was also my backcountry pilot and guide for several of those trips. Glen has lived his whole life in a remote off-the-road-system village in Bristol Bay: a village named after his grandfather. Having flown and guided in the area his entire adult life as well as raising five children there, he was also a wealth of knowledge. Often as I flew with him over vast swaths of landscape trying to learn more about the region, I would look down at some lake or river, or even a smaller pond or stream, and ask him about it. I don't think I ever found one that he couldn't tell me something about. It became a challenge to try to stump him. I always failed. Even more meaningful to me than all his wealth of knowledge is that Glen became a good and cherished friend. (I'm delighted he was willing to contribute the photographs for this book. Although he comes from a family of bush pilots who are deservedly famous for their flying, Glen's photography is equally deserving of fame.)

I've also headed out on many fishing trips with Glen that—like the characters in Van Morrison's song—have been impacted by the weather. If not rain, then wind or clouds. Although, unlike Van Morrison's characters, I can guarantee that if the sun came out and the wind died down, neither Glen nor I would ever have suggested abandoning our fishing rods to go swimming. Which brings me to September of 2022 and my final trip to Bristol Bay to complete this book. Although much of the research and learning had been completed by then, and I was mostly just looking to spend time writing *in the place* I was writing about, I would have enjoyed one more opportunity to spend some close personal time with some of Bristol Bay's northern Dolly Varden char. I certainly expected to. I had never been on a trip to the Farm Lodge in which I had not caught several fat and bright Dollies. But the wind and the rain said otherwise. Our favorite rivers where we had caught Dollies in the past had been blown out by floods, which had washed away the salmon eggs and salmon flesh. Other places we might still have found Dollies were made inaccessible to airplanes by high winds during my visit.

We attempted to fish one favorite river in high water. The fish were few and wading was brutal. We caught a couple smaller-than-usual rainbows and no Dollies. In another spot, we tried casting flies into a gale force wind. There, we couldn't even get our flies out where any fish might have seen them. I sure did miss spending time with Dollies, seeing them all dressed up in their autumn garb. Maybe a better theme song for the trip came from nine years after Van Morrison's tune when Led Zeppelin released "Fool in the Rain" (1979). The narrator in that song is also standing in the rain, but he's awaiting his lover, and feeling blue because the lover hasn't shown up. Eventually the song reveals that the fool standing in the rain was standing on the wrong city block. Perhaps it is a literal wrong block he is standing on: the woman has *not* abandoned him, but is just waiting in a different place? Or is it, as music critics have suggested, a metaphorical wrong block: the narrator is loving the wrong woman, who doesn't love him back?

After Glen and I waited in the rain (and wind) for a no-show from the fish that I loved, we finally left the block. Fortunately, we had another block to go to. When we gave up on the Dollies and rainbows, Glen led me and my friend Phil to what proved far and away the best day of silver salmon fishing I've had in my life, along with a couple enjoyable days chasing Arctic char and some really good grayling fishing. But I'm missing those Dollies, and wishing I could go on writing this book for several more years.

Works Cited

Behnke, Robert, *Trout and Salmon of North America*, (New York: The Free Press, 2002).

Berry, Wendell. "A Native Hill", *The Hudson Review*, 21:4 (1968-1969), pp. 601-634.

Branson, John B., *A 20th-Century Portrait of Lake Clark,* Alaska: 1900-2000, (Anchorage: U.S. Department of the Interior, 2015).

Dionne, M., Miller, K., Dodson, J., and Bernatchez, L., "MHC standing genetic variation and pathogen resistance in wild Atlantic salmon", www.ncbi.nlm.nih.gov/pmc/articles/PMC2690505/. Accessed on 8/27/2022/.

Evanoff, Karen E., ed. *Dena'ina Elnena: A Celebration: Voices of the Dena'ina*, (National Park Service, 2010).

Fausch, Kurt, *For the Love or Rivers: A Scientist's Journey*, (Corvallis: Oregon State University Press, 2015).

Grimholt, U. and Lukacs, M. "MHC class I evolution; from Northern pike to salmonids." BMC Ecol Evo 21:3 (2021). doi.org/10.1186/ s12862-020-01736-y. Accessed on 8/27/2022.

Kleinkauf, Cecilia, *Fly-Fishing for Alaska's Arctic Grayling: Sailfish of the North*, (Milwaukee, OR: Amato Books, 2010).

Lewis, C.S. "The Weight of Glory", in *The Weight of Glory: And Other Addresses* (San Francisco: HarperOne, 2001), pp. 25-46. Preached originally as a sermon in the Church of St Mary the Virgin, Oxford, on June 8, 1942.

—*The Pilgrim's Regress*, (Grand Rapids: Eerdmans, 2014). First published in 1933.

Schindler, Daniel, "Mosaic—the Salmon Wilderness of Bristol Bay, Alaska", Cinematography and editing by Jason Ching,vimeo. com/637271167. Accessed on August 25, 2022.

Swanson, H., Kidd, K., Babaluk, J., Wastle, R., Yang, P., Halden, N., and Reist, J. "Anadromy in Arctic populations of lake trout (*Salvelinus namaycush*): otolith microchemistry, stable isotopes, and comparisons with Arctic char (*Salvelinus alpinus*)", *Canadian Journal of Fisheries and Aquatic Sciences*, 22 (April, 2010). https://doi.org/10.1139/F10-022. Accessed on 21 October, 2022.

About the Author

Matthew Dickerson's first book, *The Finnsburg Encounter*, was a work of medieval historical fiction based on a fragment of the Old English poem *Beowulf.* Published in the U.S in 1991, it was later translated and published in Germany with the title *Licht über Friesland.* Although in the years following this publication Dickerson continued to work primarily on fiction (along with an occasional magazine article), because of his interest in outdoor sports and especially fly fishing he was invited in 1997 to pen a regular outdoors column in the sports pages of his local newspaper, the *Addison Independent.*

Finding that he enjoyed writing about trout, rivers, and ecology, Dickerson began writing freelance for various fly-fishing and outdoor magazines also, including *The Drake, Eastern Fly Fishing, Northwest Fly Fishing, American Fly Fishing,* and *Backcountry Journal.* His essays about nature and ecology in the creative narrative non-fiction genre also appeared in such journals as *Written River, Books and Culture, Creation Care, The Other Journal,* and *Christian History Magazine.*

Matthew eventually wove together the ecological thread and the trout and fly-fishing threads in several books. In 2012, he and his friend David O'Hara were awarded the Spring Creek Project Environmental Nonfiction Writing Residence at Oregon State University's Cabin at Shotpouch Creek. The result of that residency was Dickerson's first book-length collection of creative non-fiction, a collaboration with O'Hara titled *Downstream: Reflections on Brook Trout, Fly Fishing, and the Waters of Appalachia* (Cascade Press, 2014). The residency also led indirectly to the Heartstreams series of books about rivers, trout, and fly fishing published by Wings Press of San Antonio. This series began with *Trout in the Desert: on Fly Fishing, Human Habits, and the Cold Waters of the Arid Southwest* (2015) and continued with *A Tale of Three Rivers: of Wooly Buggers, Bowling Balls, Cigarette Butts and the Future of Appalachian Brook Trout* (2018), both of which contained original prints by artists in the regions he was writing about.

Five years after the Spring Creek Project residence, Matthew was selected as Artist-in-Residence for Glacier National Park in June of 2017, for Acadia National Park in May, 2018, and for Alaska State Parks in the summer of 2022. The first two of these residencies, along with multiple trips to The Farm Lodge in Alaska, led to the publication of *The Voices of Rivers: Reflections on Places Wild and Almost Wild* (Homebound Publications, 2019). The Glacier National Park residency also led directly to the third book in his Heartstreams series: *A Fine-Spotted Trout on Corral Creek: On the Cutthroat Competition of Native Trout in the Northern Rockies* (Wings Press, 2021). *The Salvelinus, the Sockeye, and the Egg-Sucking Leech,* which was largely completed during the Alaska State Parks residency, is the fourth and final book of this series.

Meanwhile, Matthew never lost interest in writing fiction, or writing about literature. His study of medieval and fantasy literature has led to several books and book chapters especially about the writings of J.R.R. Tolkien and C.S. Lewis. Many of these works also drew on his interest in and study of environmental literature. *Following Gandalf: Epic Battles and Moral Victory in the Lord of the Rings* offered a study of the theological and moral aspects of Tolkien's writings, and was a finalist for the year's Mythopoeic Society award for best work of scholarship. It has been revised and expanded as *A Hobbit Journey: Discovering the Enchantment of J.R.R. Tolkien's Middle-earth.* This book was followed by *Ents, Elves, and Eriador: The Environmental Vision of J.R.R. Tolkien* (with Jonathan Evans), *From Homer to Harry Potter: A Handbook on Myth and Fantasy* (with David O'Hara), and *Narnia and the Fields of Arbol: the Environmental Vision of C.S. Lewis* (also with David O'Hara). His current work-in-progress about spiritual aspects of C.S. Lewis's *Narnia* stories, is scheduled to appear in Spring of 2024, and is tentatively titled *Aslan's Breath.*

Matthew's second novel of medieval historical fiction—a sequel to *The Finnsburge Encounter* titled *The Rood and the Torc: The Song of Kristinge, Son of Finn* —was published by Wings Press in 2014. Just a year later, Dickerson's first fantasy novel came into print: *The Gifted*, the first of a three-volume work collectively titled *The Daegmon War* was published in 2015 by AMG. This series was continued in *The Betrayed* (2016) and *Illengond* (2017). Dickerson's published books have also included: a biography of the late singer-songwriter Mark Heard titled *Hammers and Nails: the Life and Music of Mark Heard;* a work on the philosophy of mind, computing, and religion titled *The Mind and the Machine: what it Means to be Human and Why it Matters;* and a work of spiritual theology titled *Disciple Making in a Culture of Power, Comfort and Fear.*

Matthew Dickerson lives with his wife of thirty-five years on a wooded hillside in Vermont, where they work together trying to be good stewards of the land that has been in their care for the last twenty-five years. He is an amateur beekeeper and gardener, and appreciates being within a half dozen miles of three trout rivers and a several smaller cold-water streams. He teaches Computer Science at Middlebury College, where he is also an affiliate of the Environmental Studies Program, has taught several classes in the Writing Program, and spent 10+ years directing the New England Young Writers' Conference at Breadloaf. He is a member of the Chrysostom Society and the Outdoor Writers Association of America (OWAA). And more than a quarter century after his first column about fishing, his outdoors column is still running and has won the New England Newspaper and Press Association award for the best sports column of the year.

About the Photographer

Glen R. Alsworth, Jr, is a 3rd generation Alaskan bush pilot. He is deeply rooted in the Bristol Bay region having an Aleut grandmother Mary Alsworth and a grandfather "Babe" Alsworth who was a legendary aviator. Glen grew up in the remote village of Port Alsworth, named after his grandparents who homesteaded there in the early 40s, and he has lived there his entire life. He developed a love for the outdoors, fishing, and flying while he was just a child and found his passion for photography as a young adult. Most days from May through early October, Glen is in the air flying a small plane over the lakes, mountains, rivers, and landscapes of the Bristol Bay drainage. He, his wife Lelya, and their children own and operate Lake Clark Resort, based near the headquarters of Lake Clark National Park, where they offer a place for others to experience what Glen calls "the most beautiful corner of God's creation."

Wings Press was founded in 1975 by Joanie Whitebird and Joseph F. Lomax, both deceased, as "an informal association of artists and cultural mythologists dedicated to the preservation of the literature of the nation of Texas." Publisher, editor and designer from 1995 until the final Wings volume appeared in 2023, Bryce Milligan was honored to carry on and expand that mission to include the finest in American writing—meaning all of the Americas, without commercial considerations clouding the decision to publish or not to publish.

Wings Press produced over 200 multicultural books, chapbooks, ebooks, recordings and broadsides intended to enlighten the human spirit and enliven the mind. The publisher long believed that writing is a transformational art form capable of changing the world, primarily by allowing us to glimpse something of each other's souls. Good writing is innovative, insightful, and interesting, but most of all it is honest. As Bob Dylan put it, "To live outside the law, you must be honest."

From the beginning, Wings Press was committed to treating the planet as a partner. Thus the press used as much recycled material as possible, from the paper on which the books were/are printed to the boxes in which they were/are shipped.

As Robert Dana wrote in *Against the Grain,* "Small press publishing is personal publishing. In essence, it's a matter of personal vision, personal taste and courage, and personal friendships." We found that to be true, and hope it will remain so in this new age.

Bryce Milligan

Colophon

This first edition of *The Salvelinus, The Sockeye, and the Egg-Sucking Leech: Abundance and Diversity in the Bristol Bay Drainage (from the Eyes of an Angler)*, by Matthew Dickerson, has been printed on 60 pound "natural" paper containing a percentage of recycled fiber. Titles have been set in Nueva Standard type, the text in Adobe Caslon type. This book was designed by Bryce Milligan.

Wings Press titles are distributed to the trade by the
Independent Publishers Group
www.ipgbook.com
and in Europe by Gazelle
www.gazellebookservices.co.uk

Also available as an ebook.